R.V. Mikkelson

GLITCH 1048

LifPress Wilmington

Published by LIFPRESS INC., Wilmington, DE. Copyright © 2014. All rights reserved. Except for use in any review, the reproduction or utilization of this work in whole or in part in any form by any electronic, mechanical or other means, now known or hereafter invented, including xerography, photocopying and recording, or in any information storage or retrieval system, is forbidden without the written permission of the publisher.

LIFPRESS INC., Trolley Square Suite 19C, Wilmington, DE, 19806   Contact:glitch1048@lifpressinc.com

This book contains an excerpt from the forthcoming *Glitch 2201,* by R.V. Mikkelson. This excerpt has been produced for this edition only and may not reflect the final content of the forthcoming edition.

LIFPRESS INC. Print Copy ISBN: 978-0-9904788-1-2

"Remember, democracy never lasts long. It soon wastes, exhausts, and murders itself. There never was a democracy yet that did not commit suicide."

John Adams, 2nd President of the U.S.A.

1

Paul Robinson coolly raised his Glock 17 and aimed at the attacker's flat forehead. "Nobody will be taking those machines out of the polling stations by Tuesday, least of all you—"

"No," David uttered.

"Or I." Paul focused down the barrel and past the front sight. His heart raced. *Normal, the adrenaline effect.*

"Commence firing!" the range officer barked.

Paul released five rounds into the forehead of the target in rapid succession. He liked how the gun jerked up just a little. Enough to feel the power, but not enough to mar the next shot.

The forehead area and the right eye of the target had the distinct discoloration of torn and bent paper . . . like partly-hanging chads. Paul looked over to his right.

Dr. Elaine Artois' shots were off the mark. She hit air. But then again, Elaine was here for social purposes only. Dressed in jeans and a designer blouse, a paisley-patterned one that tastefully hid some curves and bulges she'd never been too proud of, she had come with her latest love interest, David. And though Elaine had always been damn smart with media, she'd never been adept with make-up, of which she'd applied a little too much this time.

He was returning Elaine a favor. Starting from his freshman year at the Penn State, she'd been a faithful mentor to him. Being two years his elder, she taught him to tighten the writing, stick to the five W's when he'd been over his head in prose and dramatic clichés. She also encouraged him to scout out meaningful stories. Since then, they'd been pretty close. Unexpectedly, a year ago they'd ended up at the same network, but at different ends of the news spectrum. She'd artfully woven her passion for news with her vocation of medicine, as the health reporter for WQHP.

Her love life, on the other hand, was a bit of a wreck, so he owed her this much. He was vetting the Australian beau, and what better place to do it than a shooting range. "You wanna test a man's mettle, put a gun in his hand," his dad used to say. Sure, he probably meant in a bar or a hostile town, but this would do as a start. Would he try to show off and shoot sideways gangsta style? Or be intimidated by the power and noise? Would he handle the gun with respect, or as a toy, aiming it at people, if even for a second?

Intentionally, he had positioned himself between the Aussie and Elaine, so David could concentrate better on the task at hand. Paul loaded again and fired, emptying several magazines.

"Cease fire!" the range officer yelled. "Unload with action open!" The eerie silence of the range was broken by a sharp "Step back from the Firing Line!" command.

Paul turned to his left and viewed David's target.

The target boasted a few holes, the newest one just below the left ear of the faceless "attacker."

He turned to David, "Better, Dave. You shaved a whisker or two off his face. Enough to stop anyone but a Navy Seal on a mission."

"Yeah, finally getting the hang of it, I suppose. It's pretty awesome." David said in his lilting Australian accent. "No, what I'd meant earlier, was whether the electoral system as a whole, is trustworthy."

"In my estimate, the U.S. electoral system is now less reliable than any in the G-Twenty, Russia excepted." Paul ejected the magazine. "I mean, here we are, spending millions every year in developing countries, helping them execute fair and transparent elections, while back home, millions of eligible voters – ex-prisoners, for example – can't vote," Paul said, ejecting the cartridge from the breach of the gun. "Imagine, nobody could ever have voted in Australia. Why, because they'd stolen a loaf of bread? In our case, it may be because some guy was carrying a little weed."

"Yeah, pretty amazing," Dave agreed. He lifted his Smith & Wesson SW1911, chosen on account of its "retro" look, and also clumsily ejected the magazine.

"You may go forward to retrieve targets!" The officer shouted.

They did, and Paul had seven hits on target versus David's two.

2

"So, Dave, aim, exhale and steadily pull the trigger next time. Should make all the difference."

David nodded.

"Hey, let's make this next round more interesting. I'll shoot with my left hand – 'southpaw' as its called here, you keep shooting with your right. Loser buys the beer."

"You're on, mate."

A few moments later the 'Commence firing' instruction rang out again.

Paul switched the gun to his left hand. He aimed the Glock barrel carefully at the 'nose' of the target at twenty-five yards and pulled the two triggers. The first released the second. A bit left, probably just above the eye. Good enough. He repeated, favoring the right side this time. Paul wondered if his dad would have taken pride in his two-handed shooting ability.

Unlike his older sister, he had rarely been on Pa's good side. From his childhood, whenever he excelled or finished an extra dose of farm work to earn a little hardscrabble praise, he'd be told, "just don't let it get to your head," and once, after he'd devised a mechanical scarecrow that shuttled back and forth on a six foot track and moved in a convincingly human manner, his dad added, "'Fore you know it, you'll be struttin' around like one of the roosters." So early on, he had learned how to be humble.

Then again he'd never meant to set the world on fire and had been, until recently, satisfied that his conversational knowledge of "highly technical" subject matters - if one could call hog bikes, guns and combine harvesters "highly technical" - had landed him at the human interest desk. The stories hit a popular vein in Central Pennsylvania, the breadbasket of the state. Nothing earth-shattering, but excerpts from life that meant something to the farmers and small retailers; stories that, he sensed, lifted their spirits. Life, to an outsider, may have looked pretty uneventful in these parts, but dramas unfolded daily outside the political arena as well. So he didn't step into the last one . . . unless invited. But the Aussie had asked and, after all, he *was* working at the news desk of one of the state capital's four major TV stations – so, if only through osmosis, he had to have some of the answers.

3

He lowered his arm slowly, deliberately. While David had emptied his first magazine, Paul was just over halfway through his own. He inhaled the cordite deeply, appreciating the aroma as if from a fine cigar.

"Cease fire," the range officer yelled. "Unload with action open!"

Paul expelled the bullet from the pistol's breach and left the breach open. He took off his earmuffs. "And the machines aren't helping, Dave. Sure, we have no more mechanical machines with their hanging punch holes, those 'chads,' yes, true, but . . ."

"Step back from the Firing Line!" the booming voice again commanded.

Elaine stepped back to a yellow line before they did. "Didn't the government address that issue with the 'Help America Vote Act', the one requiring that you be able to trace votes, and no more mechanical machines?"

"Yeah," Paul said, nodding, "But I understand it doesn't address the 'traceability of votes' issue. Each state still has the right to choose its own voting machinery. In the case of the Commonwealth, each county. If the county government chooses the modern-day equivalent of those hundred-year-old mechanical punch stands that leave 'chads,' so be it."

"No paper trail?" David asked.

"Nope, not obligatory. And machines could still be confusing. Get this - the instructions on the ballots in New York City during one of the last local elections said to mark the ovals *above* the candidate's name. But the machine was reading the ovals below the candidate's name. Ding! Mis-vote there."

"After all this time? Haven't they had ten years to fix the problem?" Dave looked down the range, squinting at the black outline on paper.

"Maybe that's not enough time." Paul shrugged his shoulders.

"You may go forward to retrieve targets!" The officer shouted.

"Let's see who's buying the beer," Paul said.

Each brought in their targets. David's had four clean shots in the head. Paul had scored four clean, and a fifth on the jaw line.

"Good shooting, Dave . . . but you're still buying the beers."

4

"The jaw counts? I mean, he wouldn't need his jaw, he'd be shooting, not talking."

"Hmm." Paul smiled. "Good point, but still, technically it's a hit. Don't worry, I'll drink slowly."

Dave laughed. "No worries mate, just havin' a go up your nose."

Paul looked up at the glass-enclosed gallery. Latanya was back. She waved at him.

"Still shooting?" he could read her lips saying.

She was wearing blue jeans and a fluffy turtleneck sweater, one of his favorite combinations. But then again, she never seemed to wear bad ones.

She joined them near the exit. "So how'd it go?"

"Just finished. It was the final shoot-out. I was shooting with my right hand and he was shooting with his left," Dave responded. "And still, I lost."

"Against Harrisburg's own Wyatt Earp. You shouldn't feel too bad, he practices a lot. But what, no automatics this time?"

David shook his head.

"Not this time," Paul added.

"Good thing you didn't, David. Last time Paul did that, it kicked up and hit him in the head," Latanya said.

"Yeah first time with an AK-47 and had a little recoil malfunction."

"Little recoil? Honey, you had a black eye for a week! She turned to Elaine and David. "Didn't shave for five days, his cheek was hurting so bad . . . and with that patch on, he looked like Blackbeard's ghost." Then, endearingly, she looked back at him. "Well, a slightly balding ghost."

When the laughter subsided, Paul gazed at Latanya's wide smile and thought to himself, he'd made quite a catch. Sure, as a thirty-something reporter, even middling on the scale, he had his choice of Harrisburg's sexiest women. But Latanya Forrester, or "Tan" as he called her, was smart, sassy, and seemed to know his thoughts better than he did. It didn't hurt that she had super-cute dimples, a beautiful mocha complexion and a body that didn't stop.

5

He turned to David, "Next time, and we'll use Heckler & Koch's. They don't kick."

Elaine straightened her blouse. "Shall we go somewhere that serves strong orange juice diluted with alcohol?"

"So that's what you fine sheilas drink out here." David deduced.

Latanya nodded. "When we're not diluting strong tomato juice, celery stalk and hot sauce with the same."

As they walked back to the parking lot, Paul took Elaine aside.

"He's a good egg, Elaine, from what I can sense. Not that I'm a great judge of Aussies, but he appears to be real."

"Thanks Paul. Appreciate it. You know how tough it is. Sometimes I feel I'd be better off somewhere else . . . like China."

Paul chuckled, "Naw, stick around, we need you out here."

2

"They're dedicated enough, but they need a firm hand, and like you'd say, Arnie, Henderson's sicker than a dog passing peach pits," the man on the other end of the line recounted in his deliberate manner. "The other Counties are in the bag, we have good operatives there—Sylvester as well—But about Dauphin, I'm nervous, Arnold, you know the situation there is dif-"

"Ah'll handle Dauphin myself," Jenkins interrupted. "Eighty pe'cent of the storage sites are within spittin' distance a' here. Ah'll grab three hours a' sleep before the second phase starts and heck, I know the team members, 'cuz I trained them myself," he said in his homey Texan drawl.

"But you know one of the volunteers was rated high risk for dependability."

"I drew up the plan. If I can't execute it myself, what kind 'a leader am I?" he said calmly, as he flexed his biceps, once again grown almost as sinewy and powerful, after the three months of workouts, as they had been during his 'tactical consulting' service. His new responsibility had re-energized him, made him feel useful again.

There was a brief pause on the other line. "I leave it to you, Arnie."

Arnold Jenkins, thirty-four, with neatly groomed blonde hair and a white-toothed smile to match, hung up the phone. He felt far from relaxed, despite what he had tried to portray. Too much depended on so many people doing their jobs, pulling their assigned strings at just the right moment. He didn't like trusting others, at least not before he had gotten to know them personally for at least a few weeks, and not when the biggest project of his life—the right future of the country—depended on these strangers. *Too many of 'em are 'soft.''*

He walked to the bathroom and closed the hastily-repainted door behind him. He removed a pair of latex gloves from his jacket pocket and put them on, snapping them as they stretched. He lowered the toilet seat cover and stood on it. From behind the circa-1950 water tank, he pulled out a hard plastic case. He produced a small key from his pants pocket and opened the box, pulling out a compact Smith & Wesson .22 caliber automatic. The gun he slid into the holster under his left arm. He quickly snapped up the Halo X knife – an army ops version of a switchblade— obtained the year before from an arms dealer known by the name of "Brother Harrold." He closed and locked the box and returned it to its place.

He climbed back down, flushed the toilet, and from his other pocket pulled out a silencer. He deftly screwed the stubby unit into the gun and felt along the top of the barrel. The gap between the barrel and the silencer was about a quarter-of-an-inch. The silencer wouldn't snag on anything as he pulled it out. *Good.* He re-holstered the gun. *Just in case.*

Jenkins watched Ted Heinemann, the fat guy who'd looked asleep in the car step briskly to the first voting machine and painstakingly re-wrap fresh pink tape around it, precisely where the previous tape had been. He then took the plastic seal and with the same modified jeweler's screwdriver he'd used to open it, he looped and 'locked' the tapered end into the plastic holder.

Pretty good with the screwdriver, even with those cigar fingers of his.

Jenkins then reapplied the adhesive seal he had taken off earlier after a half-minute application of the heat gun.

He looked at his watch. "Better, boys. That one took two minutes, fifty-one seconds. Let's see if we can't get that down to two minutes. The groundskeeper will be waken' up in just over four hours." Once again he looked at his watch. It read 12:11. He reached to his armpit and felt the hard right edge of the gun grip. He felt reassured, in the unlikely event that some overly-curious citizen of this town ambled by.

Two hours, twenty-six minutes and seventeen seconds later, John Smith—the smaller of the two "Key Changers"—the one with the harelip, put on his down jacket, gathered up balls of discarded pink tape and let himself out of the church basement.

As Jenkins quietly closed and locked the heavy door, he gratefully breathed in the cold air, exuding pine scent from the trees lining the churchyard. As he approached the car, Jenkins glanced back and up at the dark neo-gothic structure, satisfied. *Now it's up to the voters.* He smiled to himself. *Sure, they expressed themselves in 2008, and look where it got us. Defense cuts, withdrawal from Iraq, Obamacare.* He'd never read the Affordable Health Care Act, but he didn't need to. Anything coming from that side of the aisle was trash. Like his granddad used to say: "Garbage since Franklin D. Roosevelt got elected."

He dropped off the two volunteers a block from their hotel. But as the fat one opened the door, Jenkins whipped his head around.

"And remember, we are a tight-lipped circle. Not a word to anyone," he said, placing his finger to his lips.

"Damn right," the rotund one replied. He stuck his hand up for a high-five.

Jenkins looked at him sternly and nodded.

He quietly put down his hand and saluted with a cocked gun goodbye.

John just nodded, smiling. The smile came out oddly, on account of the harelip. He looked like something crazy Uncle Ed would paint, during one of his 'Picasso' moments.

The Harrisburg branch office of the Corn Cob Caucus smelled of hot coffee, popcorn and pulsed with the adrenaline of type "A" personalities preparing to skewer a political adversary. Jenkins leaned back in the Aeron chair as he took the phone from his assistant. He crossed his left leg over his right, listening intently as he beat out a steady rhythm on his dark grey Niconda cowboy boots with his polished black ceramic pen. He put the disposable clamshell cellphone to his ear.

"They started talking. Heinemann is getting excited. He's in a bar down the street from his hotel, the Schiphol."

"What did he say, exactly?"

"He told some people they wouldn't believe what he was doing."

"Did he re-veal anythin'?" Jenkins asked calmly.

"No, but he's downing the beers fast, it could just be a matter of minutes."

9

"I'm on it." Jenkins closed the phone and sighed. He was both irritated and pleased. Irritated, because he had given a stern warning to the two volunteers, which they hadn't observed. But pleased he had suggested putting the concealed mikes onto the guys' jackets.

Jenkins pulled out the second of three cell phones he had been issued. He called the older of the two volunteers, inviting him to a celebration of the caucus, just up the road.

Fifteen minutes later he picked them up. As the pair got in the car, Jenkins removed two cans of beer from the CD compartment and handed one to each of them. "Here guys, how about a congratulatory brew."

"Don't mind if I do," said chubby Ted, his eyes unfocused, as he grabbed his can.

"Yeah," added John, the harelip.

"So, meant to ask ya earlier, whatcha guys hunt up North?" Jenkins asked.

"Moose n' deer, mostly," said John.

"Tell me about it," Jenkins started, guiding the man into conversation. Soon, they were swapping stories.

"Ever been to a titty bar?" Jenkins asked, after the hunting conversation went tepid.

"Sure," Ted replied, nodding.

"Yeah, but if it's from Northern Appalachia, you haven't seen shit. We'll show you what makes Harrisburg tick and Penthouse too.

"Penthouse?"

"Oh sure, whaddya think them little Pets are doin' when they ain't posing right and left. Like I said, y'all and those other crews deserve a little pre-celebration fer your fine 'pre-work.'"

In the mirror, Jenkins saw Ted, the older "volunteer" glance at the John, who shrugged his shoulders and smiled.

"Okay," Ted said, following up with an ear-to-ear grin, which Jenkins found truly irritating. *Why couldn't the guy get his gut stapled, save other people the discomfort of lookin' at his fat face.*

Ten minutes later, Jenkins turned the car to the left and, after passing a few closed stores and one tattoo parlor, the car approached a black sign with very bright, small letters. "The Red Velvet Pole," the sign read. In the lower part, the words 'Your Secrets Stay Here,' blinked on and off. Jenkins drove past it and turned into a darkened, narrow street. He passed an alleyway to the right, leading to the club, and parked the car.

Everyone got out.

A slight gust of wind propelled some hamburger wrappers along the cracked brick siding of an adjacent building.

"Hey guys, before we go in, would you wanna see the spot where Kerry Collins lost his virginity?" Jenkins asked.

"You mean Penn State quarterback Collins?" John queried.

"None other," Jenkins retorted.

"Sure it was here?" Ted asked, before hiccupping.

"Damn sure was. On a hot summer night. Follow me," Jenkins added.

"Cool!" the young one said.

They passed a dumpster, which smelled of mothballs.

Ted was grinning when Jenkins glanced back at the pair sauntering right behind him. One pole lamp was twenty yards behind them. The next was about thirty in front, atop the rear exit of the strip joint.

I've worked so hard helping set "The Program" up. This experiment that could let the right people, God's people, run this country the way it should be run. Business, progress, law, order, like John Galt would've wanted it to be.

Jenkins breathed in the cool air. It felt soothing to his nostrils. Calming. Suddenly he sensed the adrenaline rush, the tingling in his spine and cheeks, just like before his "vigilante" extermination runs in Iraq; well-planned events that resulted in clean, bloody executions, over quick enough to fit in a comic highlights debrief at the base afterwards over a couple of cold brews.

It's dark enough here. He pushed the button on his Halo 3. In one smooth movement, he pivoted on his right foot and swung his right arm backwards in an arc. Jenkins powered the knife quickly through half of the fat one's thorax. The blood streamed out but he had avoided the main artery. He didn't want any red splatter on his jacket – he'd had it dry cleaned just two days earlier.

11

Before the guy could even start gasping for air, Jenkins hopped to his right and plunged the knife deep into Harelip's solar plexus.

The young man, eyes wide open, grabbed Jenkins' knife-wielding hand in both of his. John was not a big man, but his arms were country-strong. The knife blade slowly turned towards Jenkins' chest.

Jenkins crouched down. As soon as the weight of his adversary shifted downward, he effortlessly flipped the man over him. The guy's back slammed the concrete like a dropped bag of cement. Gripping the knife handle alone now, Jenkins fell on one knee next to him. He plunged the knife to the right of the spot in the chest, but he angled the blade to the left this time.

Harelip's eyes opened to half-dollars. He lifted both legs up a few inches, then dropped them down violently. A speck of blood appeared on his lips. His eyes stayed open.

Jenkins jerked his head up. Fatty was still pressing the right front side of his neck in a futile attempt to stem the internal blood flow. He started wheezing. A pathetic attempt at a cough as the blood filled his lungs. *Another fifteen seconds to expire . . . don't waste time or energy acceleratin' it.*

Instead, Jenkins pulled two latex gloves from his coat and put them on. He pulled out from his pocket a plastic bag with the white powder and opened it. He bent down, sprinkling a little of it on the back of the Harelip's hand, just on the right side of the 'V' between the thumb and forefinger.

Fatty finally dropped to his knees. He appeared puzzled. His eyes lost focus. He collapsed nose down, with a thud. *Ouch.*

Jenkins stepped to his right, sprinkled a little more on his left latex glove and pressed the fresh corpse's right shoulder with his left hand, as if steadying him for a punch. He placed the plastic bag between the fresh corpse's index finger and thumb and pressed the two together.

He slipped the full bag into Harelip's front right pocket. Deftly, he reached into his back pocket, pulled out the wallet and slid it into his jacket.

It didn't have to be perfect, he was told, just confusing. Jenkin's superior officer's mantra had been '*The effort put into recon is*

12

seldom wasted,' and he once again appreciated the wisdom of that simple statement. These guys came from rough territory. Drug-related violence was rare, but brutal when it happened. And, as "lone wolves" in their communities, they wouldn't be sorely missed either. All told, this scene would come as a shock to few.

Jenkins brushed off his jacket, pleased to see he had avoided getting splatter on it. "Sorry guys, I lied. Kerry Collins probably never came here. But thanks for your cooperation anyways." He glanced briefly at the two bodies, turned around and started walking back towards the car.

He heard the back door of the club opening. Instinctively, he jumped behind a dumpster. After a few seconds he peered out from around the corner of the cold metal container.

A man stumbled out of the club, holding someone's hand. A bare-legged blonde, fur coat draped over her shoulders, followed him out.

"Only a blow-job," she said, "that's what you paid for."

"Sure, sure," the guy replied, nodding, pulling her towards the other end of the alley.

They hadn't noticed the two bodies lying fifty feet away.

Jenkins stood up and kept walking briskly out the alley. He passed three more dumpsters and threw the cell phone, with which he had called the fat one, into the fourth. Jenkins passed the last bin and turned right onto the street. *No one. Good.* He'd been lucky.

Then again, it was 12:21 Monday morning and no one, besides strippers relieving clientele of what burdened them most, would be walking the streets at that hour in this town.

13

3

"Wait, there you are!" Latanya lunged for the remote and turned up the volume.

Paul was on the TV screen, holding a microphone. "It looks like once again, voters from Pittsburgh and Philadelphia are pulling Pennsylvania's senatorial contest into the blue camp, in spite of early red leads for house seats. The exit polls show that Democratic U.S. Senator Duquesne is poised to take the Senate seat."

The view cut to the news anchor. "That was Paul Robinson taking a poll at two o'clock this afternoon, which pointed to an early lead for Duquesne."

"Another ten seconds in my fifteen minutes of fame," Paul said, quietly. "And I'm a reporter!"

"Come on, Human Interest promoted to Politics? That's great," Latanya blurted out.

"It's not like that. It's 'Politics Lite.' Besides, they were short on staff, so all hands on deck."

"Oh don't be so self-effacing. It was a nice break, Paul, you should run with it!"

"Yeah, for Merry-Jerry? You should see where I was this morning, thanks to him. He says, you know in that hyper-excited voice he has, 'Paul, did you know that Adams County is the site of the oldest polling station in Pennsylvania?' 'Had no idea,' I answer, you know, rubbing the sleep outta my eyes. Anyway, he squints at me with those beady eyes of his and says: 'Still in existence after almost one-hundred-and-thirty years! Now isn't that something voters would want to know about?' *Sure*, I almost said, *but not nearly as much as they'd like to know the average age of the interns you're bonking.*' I just about answer him back, but, at the last minute, I restrained myself—" he gulped down the remaining ounce of Troegs ale from the bottle and put it down. "And ended up going to—what did Dave call it, 'Woop-Woop', to cover a bunch of 'Bogans' voting. Carried on about what a polling station looked like 120 years ago . . . "

"I thought Dave's definition of 'Bogans' was cute – 'hicks with a beer mug handle glued to their hands'. Definitely apt for some areas around here," Latanya said.

Paul's cellphone rang. He picked it up and looked at the screen, "Speak of the devil." He put the phone to his ear. "Yeah Jerry?"

"Hey, have we got a phenomenal lead in town right now."

"No, Jerry, no—I've done the oldest polling station, trends, the polls, the 'up close and in-your-face personals.' Hear this?" Paul grabbed a new bottle and twisted the cap off, letting it "pop" next to the phone. "That's the sound of 'I'm done for the day.'"

"Listen, Paul, this could be a *big* story, and the rest of the crew is covering the victory parties—" Watkins started, as if he had just found the largest gold vein west of the Sierra Nevada's.

Paul listened a few seconds longer, shaking his head as he looked at Latanya, "All right, all right. You win . . . I'll go. I'll leave in two," he said, not trying in the least to hide the reluctance in his voice. He turned around to see her looking at him. He sighed . . . "Another ten seconds that I'd gladly trade for another half-hour with you."

"Aw, that's sweet honey, but you've been kind of just sitting around here an awfully lot."

Paul shrugged. "Some old lady who had trouble with the machine at the polling station claims it changed her vote. The same lady, no doubt, who couldn't program her alarm system," Paul shook his hands, "or understand Skype. *Oh, there's that green circle and a red one.*"

"Maybe not. You might be proven right about the voting machines, who knows? Go ahead, I'll be fine. I'll use the opportunity to drop in on Mom. She hasn't seen me in ages."

In his navy blue Ford Fusion, Paul waited behind an old Buick for five seconds after the light turned green.

Another car drove up behind him. The driver of the front car was clearly chatting up a young brunette passenger. Another five seconds passed.

"Come on, come *on!*" he blurted out, to no one in particular. He rolled down the window. *Forget this.* He got out of the car.

15

The driver of the car in front got out at the same time. He weighed, Paul estimated, about 220 pounds, and most of that muscle.

"Hey, my good man, I'm Paul Robinson, WQHP, rushing to my next story!"

Yeah Joe, Ah got a problem with the battery," he said with what appeared to be a Texan drawl. "Ya couldn't start me up, could you?"

Paul glanced at his watch. *If he were alone, I'd make him wait, but with a date . . .* "Okay, I've got some cables, pop open the hood."

As he got in his car, two black cars, an Explorer and a long Lincoln zipped by, ignoring the red light. They turned left at the next block.

The man had, apparently, never seen jumper cables in his life, so Paul had to make all the connections. *Damn, I promised to be right there.*

As he pulled up to the polling station, the WQHP cameraman was putting his gear in the van. *What?* Paul jumped out of the car. He strode to Watkins, "So, where's that defective lady, er, defective machine?"

"We were late, Robinson. Some fender bender a few blocks from here kept us penned in for three-four minutes. So Channel 7 got here first. Pinched the story . . . if there ever was one. They took the woman, too."

Paul stared at Watkins in disbelief. He decided it best not to mention his brief Good Samaritan episode, as it made no difference. "Channel 7, hunh. Wouldn't surprise me if they set us up, they're so desperate. But look, Jerry, from my house, this voting station took—" he looked at his watch, "seventeen minutes, less even, and you were here what, two minutes ago? How could someone have beaten us to it?"

***

Latanya sipped the last half of her coffee slowly, tasting nothing, as she digested the word 's-l-a-c-k-e-r.'

"Now Latanya, honey, it's not that he's white. That don't botha me. You know yo' Dad and I ain't no racists. No reason to

16

be, in these parts. We worked hard, and we got our due. Maybe the gels and hair straighteners ain't for everyone, but da financin' we got from banks run by white folk. You remember, Mr. Winkleman?"

Latanya turned her head to the side, mouthing the words as her mom uttered them. That she'd heard this ten, if not twenty times before, had no effect on the onslaught. It always started this way, never abridged, before moving on to the same conclusion every time. Yes, her parents had grown their hair gel and grooming supplies business into a statewide juggernaut, servicing a network of 2300 stores . . . and all this from a mere four-page business plan and a presentation to a savvy bank loan officer who hadn't an ounce of racial prejudice in his body. But how that commercial success provided the justification for lecturing on any topic under the sun, from space exploration to dating – which she sensed was the next topic on her Mom's agenda – escaped Latanya. She braced herself.

"You're twenty-seven. You ain't gonna be young foreva'. Soon after you get yo' 'P' 'H' 'D', you gonna want kids—"

"But Momma, I don't want them right now, okay?!"

"I understand, child . . . but you will. And when you do . . ." Mrs. Forrester stood, waving a silver-plated teaspoon at Latanya, "you're gonna want a real man next to you. One who, at thirty-two, ain't tryin' to race cars, or climb mountains, or change jobs at the drop of a hat."

"But Momma, he's just . . . he's just not like the others."

"What? you can't find no others like him, what with yo' Halle Berry looks?" She moved to the kitchen window and looked out into the pitch black outside. I believe I can see a line o' men stretchin' from here to Lancasta', willin' to give you the keys to their fancy cars just to date you, girl! Why the Jefferson's son, he's a successful Philadelphia lawyer . . . "

Latanya glanced at her State junior kick-boxing championship trophy, amongst other mementoes of two successful—and one not-so-much—children in the glass cabinet in the dining room. "Yeah, yeah, Ma, I gotta run," she interrupted, grabbing her mug.

Several drops of coffee spilled onto the floor. She clanked the mug back down on the dark green granite counter top and

17

stormed to the kitchen sink, avoiding her mother's gaze. She ripped a paper towel from the burnished steel holder, stomped back to the spot and started wiping the coffee drops. A teardrop fell, adding to the small brown smudge still there.

4

At eight a.m. the day after the election, the pink sky enveloped clouds and skyscrapers alike. Overnight, the temperature had hit 29 degrees, but was inching up to 35 now, with the frost no longer visible on the grass growing between the cracks in the sidewalk. In a glass-enclosed circa 1950's Diner on North Second Street, three men dressed in jeans and T-shirts discussed something technical, and heatedly, as only dedicated wonks could.

Charles Duprix, hair disheveled, brown eyes darting to and fro like a cobra's, sat across from a younger Martin Leboeuf. Joshua Kendrey, the lead programmer, sat next to Leboeuf. Kendrey, green-eyed and hair tousled, appeared despondent. He probably hadn't showered in a couple of days – he exuded body odor. Duprix thought it insulting, even for a programmer.

It had been okay working with Kendrey, but the guy was a dinosaur. Smart, but old school . . . and old. Almost forty. But Kendrey had one big thing going for him; he wasn't Marty. Duprix had suspected weeks ago, that Marty Leboeuf had been hired primarily for his family's contacts with the state's government. Not only did Leboeuf look like a younger version of The Penguin, but turns out, he couldn't program worth a damn. . .

With his fingers tapping out a rhythm on his thigh and his jaw locking every ten seconds, Kendrey was testing Duprix's patience. But Duprix still needed the answer to a question that had been gnawing at him for a month now. He turned to Kendrey. "Josh, why did you get to do the core code?"

"Because I'm . . . sex, sexi . . . ," Kendrey stuttered, "sexier than Hugh Grant."

"A freak is what you are! If it weren't for the Xanax, a zombie," Leboeuf blurted out.

"I rep . . . I rep . . . resent that remark!" Kendrey tore out, before his jaw stiffened, revealing a good angular cut. He could speak no more.

"He may be a freak, Leboeuf, but you . . ." Duprix blurted out, pointing his fork at the early-balding man, "didn't check with either of us before finalizing the last chain. That's *so* uncool." He ignored Leboeuf's glare as he turned back to Kendrey. "Look Josh, just tell me again that there were no major fuck-ups. You're the lead. I do totally not want to kiss my three hundred K bonus goodbye. Or my ride, which I'd have to sell if someone finds out that we couldn't program these things right, because I'd never get another programming job in my life."

"No . . . Drive on. One little problem with seg-, segment two 'D' on the last chain," Kendrey said, calmly." Marty kind of took a shortcut on the s-s-string for rejection and acceptance . . . "

His mind wandered as he bent his head down and looked out the window. The house across the street had these really neat copper-framed bay windows. And behind one of them on the second floor, were two men, one with a massive camera. *Photographing the diner? Why the diner?*

"No way! Don't listen, Charlie, this mutant's foaming at the mouth again," Leboeuf corrected.

"Marty, for the condition to test, when the value of the user action variable is more than one, if the value m- m-matches the expression, the code following the case line is executed, otherwise it acti-. . . activates a default," Kendrey sputtered out. "But shouldn't be a b-big p-problem."

"Duh," responded Leboeuf. "I can't believe we're even talking about this. They've tested the machines, what, three dozen times? Nothing. No issues. Everything is just fine. John Q. Public is perfectly served. If some folks are so old they can't deal with 'D'-'R'-'E' machines, maybe they shouldn't be allowed to vote. They sure as hell can't drive to a polling station, 'cuz that takes about four million more neural reactions."

Duprix rubbed his tousled hair with his left hand as he cut the last two of the short-stacks, soggy with maple syrup, in half with his fork. "You Danabold clowns can do what you want, but if the program is really fucked up, I'm gonna drop the word at Harker and fast. Whatever happens, I'm keepin' my ride," he concluded before stuffing half of the stack into his open mouth.

Leboeuf stared at the wall ahead.

20

Kendrey looked out the window. "How fast does it go?"

"A hundred so far . . .*chomp, chomp* . . . and, I haven't pushed it at all," he answered, squeezing words out past the clump of food in his mouth.

Kendrey and Leboeuf looked at him, wide-eyed.

"Where can you drive it that fast?" Leboeuf asked.

He swallowed his mouthful. "Besides the racetrack, not many places. But I like this stretch from Carsonville to Lyken just north a' here. Even before that, for about seven miles the speed limit's high for winding hills and a narrow road. But from Carsonville onwards, it's *really* sweet. At mile three, no more houses. I pop that thing into top gear," he said, moving his hand in the gesture of a gear change, his eyes focusing with a steely intensity. "A mile stretch, hilly but straight. Mile four, a ninety-degree turn to the left. Chck-chck *Er-Ruh*. At that point I'm at the sign that reads, 'Welcome, State Game Lands—Hunt Safely.' And 'yeah,' I say, 'Ready to hunt for my hundred '*M*'*P*'*H*' once again." Duprix smiled maliciously. "As good as the Nurbergring track in Bavaria, with those trees whizzing by. In fact . . . " he drove his fork into the remaining half-stack of pancakes and simulated a stick-shift. "I'm going for a *hwurrh*-chik-*urrh* drive this afternoon, you guys wanna come?"

"No way. You know the rules! We, we shouldn't even be meeting here like this!" Leboeuf lashed out.

"Yeah, yeah. I thought you'd chicken out. When's your wife gonna give back your balls, or is she still using 'em to soften butter?"

Kendrey joined in a chuckle.

"So funny, I forgot to laugh. You guys, no responsibilities in life . . . should just be lined up and shot," Leboeuf said, as he once again rolled the edge of his napkin into a tight little tube.

"What about you, Josh?" Duprix asked.

"No, that's okay. Too much ex-excitement for this d-dude."

"You wimps don't know what you're missing." He stuffed the last fist-sized combination of pancakes, flavored corn syrup and 'natural' Canadian bacon into his mouth.

***

"So, what's the scoop with our drug mules?" Detective Harry Olson asked his gray-whiskered, narrow-eyed assistant Fred Agnelli. He lit a cigarette and sucked in deep.

"Our man in Coudersport reported that Smith had told his grandmother he'd volunteered to do some campaign work for the Republican Party. The Harrisburg branch confirmed that the two were on the list of volunteers for the get-out-the-vote push before the election, but never showed up. But the phone numbers of the Party representative working the phones didn't match up with the number Heinemann had called —and from which someone had called him about thirty to sixty minutes before T-O-D."

"Yeah, just an alibi. Weapons match-up?"

"Friday."

"What, backlogged again?"

"Yeah, the three hunting accidents hadn't been cleared up yet. You know how these hunters take an arsenal with them nowadays. Landsmann has to test eleven rifles, and that's from four guys."

"Yeah, fine. What about the drug angle?" Olson took another drag from the cigarette.

"Nothing. Finnegan's meeting with the last group of dealers, but no one's seen these guys. Then again, they never do."

Olson looked out the window, shaking his head. "Why didn't they go through Philly? Like Harrisburg is a big drug pipeline. Unless . . . Is something new coming down the pike, Agnelli?"

"Not that we know of. But I'll find out from the DEA tomorra."

"Keep on it. We've been labeled fat, dumb, and happy once before, and I don't wanna go through that experience again, guys."

***

At about sixty square feet, Paul's office at WQHP was the next small step up from a cubicle, but symbolically, the few extra feet and more importantly, the walls and door made it, relatively-speaking, a palatial chamber compared to a lackey's space on a stone basement floor.

While he waited for the "Pennsylvania Votes" website to download, he replayed last night's dialogue with Latanya in his

22

head. "No comforting for the scooped reporter?" he had asked in his lowest tone, when Latanya had hinted she wouldn't be staying over. She had seemed distant, stressed about something.

He refocused on the screen. Huntingdon County . . . used the optical scanners. But in the areas of his South Central Penn beat, Dauphin, Blair, Bedford, Berk counties were still using technology that left no paper trails.

Mazurski entered his office.

"History was just made, and you're, what? shopping for an X-Box?"

"Yeah, wanna play Halo while I wait for my assignment. "

"How cool, who wouldn't?"

"Actually, I'm looking up the machines that were used here in the last elections."

"Hunh, what for? The ones that broke down or were hard to use didn't receive the State Department's sign-off, did they?" Mazurski asked, with his earthy South Central accent.

"Could be. I dunno what machines they're using in each county. I mean the two-step optical scanners are fine, the ones where you vote electronically—then get a print-out that you stick . . ." he gestured with a horizontal slide of an invisible sheet, "into another machine that scans it. But the other electronic machines? Look, I just saw here on Google, that last summer, some guy hacked Pac-Man into one of the voting machines. And that was a . . ." he twitched two fingers to make quotes, "tamper-proof" machine. Nice surprise for any citizen who would've used that thing: 'Oh, I hadn't planned to, but turns out I voted for Pac-Man, you know, from the Eat-or-Deflate Party.'"

Mazurski chuckled. "Sounds like a great practical joke to me. Besides, what would you have, a return to hand-counting?"

"Yeah! People's thumbs should look like they stuck 'em in Grandma's blueberry pie for an hour."

"Uh-hunh, medieval democracy, Afghan-style," Mazurski declared.

"Dry, tasteless goat anyone?"

"So Paul, whatcha think you'll be doing next?"

"My prediction? The zoo. It fits like a glove. Besides, I hear they've trained the chimps to hold up signs at anti-abortion rallies, the kind newly-elected Dooley wants to organize."

"Paul, WQHP viewer ratings were in the toilet this month. I'm not sure they'll run with any more zoo stories to revive 'em, either. Rumor has it they're letting people go."

"Serves 'em right. I suggested we do a story on how raising the minimum wage would reinvigorate the large pockets of the country not lifted by the Shale play. But *no*, 'too risky.' But now, I'm going to explore something else, cleared by Merry-Jerry himself. I suspect that these machines haven't been tested enough with retirees. Besides, he even hinted that the story about the lady at the voting booth the other night might be 'big.'"

"Look Paul," Mazurski replied, crossing his arms over his chest, "you're pretty funny and small-town clean. But you know who owns this network. As a friend, Paul . . . just be careful."

5

The foothills of the Berry Mountains rise abruptly out of the valley. Around 10,000 B.C., the Iroquois Indians lived here and called this collection of ash, elm and other hardwoods interspersed with soft pine trees, small creeks and abundant wildlife, the land "Where we fed on eels" or 'Swatara.' It was also the name of a State Park twelve miles due East of where Francis Antoine was now overseeing his latest 'job.'

He sat in the front passenger's seat of the Lincoln sedan, admiring the stately old trees, soon to be his accomplices, if events evolved as he had planned. He bore a scar on the side of his neck from a decade-past encounter with a knife-wielding security guard, and another horizontal one above his left eye from the 9mm slug that grazed his head in the Bolluci encounter. Antoine wore a leather jacket, custom fit for his undersized frame in Italy, and Moschino sunglasses that glistened when specks of sun occasionally burst through the clouds and streamed through the few remaining forest leaves. But today was primarily a day of thick gray overcast, the kind spread with a trowel, not a paintbrush.

Antoine had sized his client, sitting in the back seat, up pretty quickly. The guy didn't go out into hunting country in November . . . or any other month of the year for that matter. And Antoine sensed he wouldn't be invited to sip Glenlivet with him either, even though the man had called him only four hours ago with this job. In this biz, that was beyond a rush job—it was to deliver a miracle. *Four frickin' hours.* If the pay wasn't so good, he'd have sent him to hell.

And Antoine rationalized that he'd worked for worse characters. Much worse, he reflected, as he watched the two road workers throwing some dirt onto the empty spot on the embankment. *That's not it.* Antoine jumped out of the sedan and bounded towards Tom, who was lamely depositing another shovelful of dirt into the hole.

"Come on, come on guys. Tom, pile it on. Move it!!"

The small half-boulder stood in the center of the right-hand lane. It looked, from Antoine's vantage point, as if the Creator had whacked a three-foot wide gray tennis ball, embedding two-thirds of it on this curvy stretch of back-country road four-hundred-thousand miles below.

He walked twenty yards towards the pick-up. "Okay fellas. Parkett, take ya shovel and tree branch trimmer and come here," he calmly ordered, in his thick South Philly Italian accent.

The tall, broad-shouldered man, in late-Fall roadwork gear obediently took his shovel and stepped quickly towards him.

"Now find some smaller rocks—ya see up there on the ledge?" he said, pointing to some ten-pound stones that had been polished with the ages, to a shiny gray-black sheen. "Take the branch trimmer and use it to knock two of 'em down and make it look like they rolled here. Got it?" he asked, gruffly, not expecting an answer.

Parkett nodded.

Yeah, sure he got it. Parkett was good with arms, but he had the IQ of a side of beef. But that was okay, too. At the end, Antoine needed his aim, his muscle, and his silence.

"And *fellas*," he said, turning to the other 'roadworker' on his left, straining every muscle in his neck not to sound condescending, "make it neat, will ya? This part of the road's gotta be cleaner than the floor of the Vatican."

The men nodded. "Gotcha, boss," Vince replied.

He looked towards the third hire, a guy of medium build who was slow with the shovel. "Tom, put the shovel down and take a couple of water bottles from the back of the truck. Use that pipe freezing machine there, like I showed you, and freeze the water," he barked.

Seconds later, the generator started up with a deafening roar.

"Vince!" Antoine shouted to the second worker, standing by the generator. Antoine had used him before, when he had to whack someone on a busy road. A second cousin, the guy needed the money—he hadn't figured his retirement plan out until six months before he was retired—but he wouldn't squeal, he could be trusted, and at five grand, was a deal. "Before we go at the end, ya put this boulder back where it was. Then cover it with leaves,

26

like nothing ever happened. Just leave the two small stones on the road. Okay?"

Vince nodded.

***

Jenkins, dressed in a hunting jacket and boots, appeared to be having a late lunch at a window table of the Carsonville store and diner. He looked outside, over his beef brisket, as a Porsche, just a shade yellower than his corncob side dish, roared past. Jenkins heard the deceleration of the explosive engine, as the driver prepared to take the 45-degree curve. Jenkins calmly picked up the cell phone lying on the table and speed-dialed a number. "The buck're hoppin' today, Frank," he mumbled.

***

The sound of the portable generator pierced the air. From his third two-quart bottle, Tom splashed a pint more water into the oversized plastic disc. He plugged both ends of the jerry-rigged pipe freezing device into the water and like the previous splashes, it started stiffening to ice in less than thirty seconds. A narrow sheet of slushy ice now extended from the boulder in the right lane to the center of the opposite lane.

As Antoine walked back towards the car, he tapped Vince on the shoulder.

The worker turned to him.

With his oversized hand, Antoine made a chopping sign on his forearm.

Vince nodded, turned to Parkett and called for him to stop. Then Vince took their shovels and buckets. Parkett picked up the generator. Both followed Antoine.

Antoine turned to Vince. "Now back the pick-up into dat gravel road at the bottom of this straight stretch –down there . . . " he pointed. "Back it up deep. Nobody needs to see nothin'."

Antoine stopped and surveyed the scene. He then climbed into the sedan. He watched as the pickup reached the gravel access road and disappeared from sight. He signaled and the driver drove the sedan just past the same gravel road and backed

27

the car in front of the truck. From the road thirty yards up and beyond, the vehicles wouldn't be visible.

Antoine put on a hunting jacket and climbed out of the car. He looked at his Fendi watch, and heard the faint roar of a powerful engine. He started jogging up the road, which soon became a mere trail, running parallel to the paved road. He heard the engine again and broke into a sprint. He stepped off after thirty yards and into a copse of trees between the trail and the road. He found himself just above the curve, with a reasonably clear view of both the boulder to his right, and the gently curving stretch of road to his left.

He didn't need to strain anymore to hear the 2.7 liter engine as this other-worldly sports car powered to within two hundred yards of the curve. He saw glimpses of yellow bobbing through the forest.

The car would be there in five . . . four . . . three seconds. Antoine heard an explosive roar. *A man's car. Good choice.*

The left front headlight emerged. The full car appeared. Forty yards. A polished yellow engineering masterpiece.

Antoine adjusted his gaze to the windshield and saw the driver's face.

The programmer's eyes widened as he tried to round the last easy curve. Must've been doing 80, maybe 90. *Fast. Bravo.* The man's hands jerked the steering wheel a quarter turn to the left. *He saw it.*

The front left tire hit the slushy water. Brake pads gripped the wheel, screeching. The turn of the car accelerated.

The scene switched to slow motion as the yellow Porsche 911 levitated over the guard rail, spinning into the trees behind it. The nearest tree stood fifty feet along the trajectory the car had taken from the road. *This must be the precise moment, during which his life flashes by him. Every major event, every memorable interaction with his siblings, his parents . . .*

The vehicle first hit the sturdy ash with the protruding wheel well that covered the rear right tire. The impact violently spun the Porsche clockwise three times. The car continued flying forward until it hit a second tree, this time with the driver's side of the car. *Teachers, bosses . . . first girlfriend.* The top of the tree jerked back ten

28

feet. Upon impact, the old elm shed what seemed like a hundred years of dust, old leaves and insects. Metal screeched and glass shattered into thousands of pieces as the Porsche continued to wrap itself around the trunk, like a cheap tin toy car Antoine enjoyed taking from the neighbor's kid and bending with a screwdriver when he was four.

The car pushed farther and farther into the trunk, until, resembling some odd UFO attaching itself to a tree, it stopped moving entirely. An eternity of pure silence followed.

Finally, the car responded to the law of gravity, and started its vertical tumble.

Antoine took a step closer and watched the car turning over, first showing the deeply dented side, then the flat one, breaking branches small and large, before landing roof side down. Dust and twigs continued to rain down, until the car was almost completely covered.

He calmly walked down the embankment, across the road. He vaulted over the guardrail like someone years younger, and sauntered down the hill to the car.

From close up, the vehicle resembled a big yellow accordion, shaped like the one his grandfather played to make a little money in Newark thirty years earlier, but with a glass shelf jutting out incongruously, and crimson splotches on the creases. The windshield was half off the front.

After three seconds, when there was still no movement behind the wheel, Antoine reached through the missing window and felt for the pulse along the driver's bloodied, distorted upside-down neck. He felt none.

He walked in a zigzag back up the hill. He was wearing road worker standard issue steel-toed boots so no one would identify him by his footwear. As he stepped over the guardrail, he saw two short skid marks.

He looked back at the car. Barely visible. *It'll be three days at least, before weekend hunters find the wreck. Perfect.*

Vince and Parkett had exited the truck and tilted the half-boulder on its edge.

The third guy was still looking at the car.

29

Antoine fixed his gaze on him, and he promptly went about his clean-up job.

Parkett and Vince had already rolled the half-boulder back into place.

Antoine stepped to inspect Vince's work, which now involved covering the boulder with leaves, just as it had been an hour earlier.

Vince turned his head towards towards Antoine. "Hey, Frankie, I got a . . . a question."

Antoine looked at him. He didn't usually get queries from his sub-contractors, even from those like Vince, whom he'd known since childhood. He knew it wouldn't be a question of more money or some weird-shit demand, but still, Vince must've known he was taking liberties. "Yeah, sure, Vince."

"Why are you doin' this?"

Antoine's head jerked back.

Vince gazed back downwards at the little streams of water formed by the melting ice and moved a foot to the left before looking up again. "No, I mean—this guy was one a' theirs, and from what I sawr, I mean, just a kid," he murmured in the accent of the neighborhood. "Why didn't you stick to workin' for one of the families, you know, our people? At least you know if they gotta whack someone, it's gonna be some turd who had it comin' to him."

Antoine's gaze met Vince' eyes when Vince looked back up. "Vince, if you was any younger, or I didn't know you, I'd belt you for askin' that." He waited until Vince was sufficiently cowed, his eyes looking back down on the ground, before continuing, "My grandpa told me ya gotta be on the side of the good guys sometimes, even if they ain't ours. These people have their problems, too. I mean maybe this guy was one of them 'identity thieves.' Maybe he stole a couple a hundred million from old folks, like your cousin Maria. *I* dunno, do *you* know?"

Vince shook his head.

"So, maybe the most important thing is just doin' our jobs da best way we can. Right?"

Vince nodded, then smiled nervously. He brought his eyes up briefly, but didn't look into Antoine's eyes, even though Antoine stood a full three inches shorter than him.

"Good Vinnie, now let's get outta here," Antoine said, patting Vincent's shoulder with his disproportionately large chop of a hand.

As Antoine walked back towards the sedan, he recalled the exact words of his grandfather, spoken forty years earlier. Gramps had just come back, bone-tired, to the tenement house apartment after playing another gig at the Florentino family's villa.

Antoine had waited up two hours past bedtime to be with Gramps.

"Frankie," he said, turning to him, "rememba' this—no guy who's made it, is clean. 'Da people I play my music fo' . . . " he started, while shaking his head, "Every one of 'em's got blood on his hands. Loads of it." He sighed deeply, turning away, looking onto the old scuffed-up refrigerator. "I thought I'd live better clean, but I just didn't have 'da brains or 'da *coglioni,*' ya know, the two between your legs." He turned back to Antoine, looking him deep in his eyes, with his own bloodshot ones, as if trying to find something. "When ya grows up, do what ya gotta do," he said with a tinge of despair, "Just rememba' three things: Ya got brains, use 'em, Second, come ova' here . . ." he paused, waiting for him to walk over, then gripping his eight-year-old shoulders firmly, "Don't eva' do any dirty work for us Guinea's. Ya can't trust our people. Do it for Anglos, 'da Kikes, the goddamn Chinks, if ya have to . . . just not for our people. You'll end up like ya' father, a bullet in the eye for a 'thank you.' *Capiche?*"

That was actually the second thing his grandfather said.

Antoine remembered he had nodded, pursing his lips tightly.

"And the last thing is, Frankie, neva' sell yourself cheap. Don't eva' let anyone walk over ya, treat ya like ya wasn't there."

He would never forget how Gramps held him that moment, tightening his thick-fingered grip on his shoulders, breathing on him with his sweet bourbon breath. It made him feel special, like the Six Million Dollar Man, but at the same time, defective at his core, a broken toy that could never be fixed.

31

Antoine opened the sedan door, sat in the passenger's seat and removed his gloves in two swift movements.

"Hell of a wreck. So, was he breathing?" his client asked.

Antoine sensed his client's doubt, his ignorance of basic human physiology, and this enervated him. "Whaddya think?"

The man smiled a nervous smile, the mealy-mouthed kind. "Not after you took care of him."

Weak. Antoine looked through his side window, away from the driver. "Yeah, Dynamo is defunct," Antoine stated, matter-of-factly. He turned towards the client and looked him squarely in the eyes. "He was already dead when I got to the car."

The man smiled again, but Antoine did not change his expression; he was not amused and watched the man's smile fade to a frown as he glanced at his watch.

"Classy wheels by-the-way. Too bad for the guy," Antoine commented.

"Great work, Francis. Let's close up shop."

Antoine didn't answer. *I say when to "close up shop," asshole.*

Parkett got into the car.

Antoine watched Vince get into the pick-up. As the driver put on his sunglasses and turned the sedan around, Antoine held up his thumb to the driver of the truck. In the rear-view mirror he watched as the vehicle turned and took off in the opposite direction, towards the diner.

The sedan slowed down as it approached the brightly-donned road worker next to the 'Road Closed' sign. Antoine lifted his chin slightly. "We're good, Jim." The back latch door clicked open. The man folded the collapsible legs of the sign and walked to the rear.

6

It was 2:30 and Paul sat in his office, fast-forwarding a video recording of another channel's news story, from 2006, about voting machines stored in an unlocked warehouse prior to elections. Mrs. Muller was still not answering her phone. Paul heard a knock on the doorframe. He turned to see his doorway blocked by Borgen's administrative assistant, holding a pink slip in one hand.

"Mr. Robinson, I was told to tell you not to take it personally."

Latanya paced back and forth in front of the couch. She stopped to chew her fingernail. Paul looked forward, staring at the TV, not to view anything particular, but to avoid her penetrating glare.

"I cannot believe it!" she bawled out for the third time.

"Tan, I'll get another job."

"For lower pay, and then get fired again. Paul, don't you understand, that's not the way it's supposed to work!" She paused long enough to draw a breath. "God, what was I thinking? A commitment!"

"I'll earn it—"

"And spend it just as fast. You buy two thousand bucks of climbing gear, and yeah, it's in the trunk of your car, but it hasn't seen the side of a cliff for what, a year?"

She stood for a moment, statue-like, before a tear fell from her left eye. When he approached, she stepped back and waved him off.

He lowered himself back to the couch.

She scrutinized his face for three seconds.

"Oh forget it. I'm going. Momma's right, like always."
Latanya stormed off, leaving him alone.

"We'll be fine, you'll see. I'll get an exclusive and then—" he called out loudly. He was answered by the sound of a coat being taken off the hook and the front door opening. *Why is she acting so strange?*

"An exclusive? Sure, when palm trees grow in Pittsburgh!"

As he got up and rounded the corner, the door slammed shut. He felt a cold sensation in his chest. *Why this fit?* After all, she'd seen him change jobs before. There were five TV stations in town and one he'd left earlier hinted they would take him back. After a few minutes, he shrugged his shoulders and punched 'Ohio for meeting with Danabold' into his iPhone calendar.

The next morning, at nine o'clock, Paul walked into the 'Café Ole' coffee shop carrying a bundle of magazines and newspapers under his left arm. He went straight to the counter.

The lithe, dark-haired cashier smiled at him.

"What brings my second favorite reporter in on a workday morning?" she asked, smiling.

"Second favorite, Frida?"

"Oh, okay. This morning, my favorite."

"That's better," he paused, sighing. "A search."

"Fired again?" the cashier frowned.

"Orwell...Vonnegut, now me."

"Did they also drink double latte's with cinnamon syrup?"

"Undoubtedly. Today I'll have a triple."

"Oh, a serious search."

"But not for a job."

He walked to a free table, dumped his newspapers on top and sat down. He opened the Patriot-News to the third page. His cellphone rang.

"When can I come to the apartment and pick up my stuff?" Latanya asked.

"Tan, don't."

"Paul...Let's just get it over with. It's not easy for me either. It's just...I'm... When will you—"

"Be there?"

"Not."

"Triple Latte! Cinnamon," the barista interrupted.

"Listen, Tan, I'll be flying this afternoon to Cleveland to interview someone, so I'll call—"

Latanya had hung up.

He slammed the phone down on the chair, stormed to the end of counter and grabbed the latte. Yes, he understood he wasn't demonstrating a typical ambition. But in the past, Tan had encouraged the rebellious streak, in all ways. He left Channel Eight because they had insisted he cover debutante balls and calf birthings. She'd supported him through that episode. What had changed so dramatically in the past year?

Hamilton Street used to be a staunchly middle class pocket of Harrisburg. But now, struggling and lower middle class African-Americans lived side-by-side with the aging Italian- and German-Americans who had built this part of the city.

He pulled the Fusion up to the curve, not far from the railroad tracks. He walked to the house. An old patchy-leafed cherry tree stood in front of it. Rust showed through parts of the metal handrails next to the steps. A bright light spot projected onto the front porch.

Paul looked up to see a dime-sized hole in the porch roof. He rang the doorbell.

From inside he heard a shrill, wobbly voice, yelling, "I'll be right there!"

He waited for twenty, thirty seconds. He tried looking through the circa-1980 peephole, but apparently they had already started making the wide-angle ones back then, so he saw nothing.

"Who is it?" he heard, followed by a hiccup. "What do you want?"

"Paul Robinson, reporter from WQHP. I'd like to talk to you about your experience with the voting machine."

"I already told it all to Channel Seven; they even showed me on T-V," Muller responded, with a shaky voice.

"Oh really? Would you have a minute to talk about it?"

Five seconds of silence was broken by, "I'm a very busy lady."

He heard a hiccup through the door.

"I won't take more than ten minutes of your time, I promise."

"Oh, all right. But let me see your journalist . . . documentation, first."

35

Paul stepped back from the door, pulled out his press pass, which WQHP hadn't yet confiscated, held it next to his face and smiled.

"Recognize me?"

He was answered by the sound of lock tumblers turning. Paul wasn't quite prepared for what he saw when Muller opened the door. Judging by the stained, ill-fitting dress and white hair cast in all directions, the tall, the tall, gaunt retiree appeared to have been serving crowd control duty after a Bruins Stanley Cup win,.

Undaunted, Paul entered, smiling.

The drawing room seemed neat enough, but the thin rays of sunlight, poking through the closed drapes, lit up a thick cover of dust on the furniture.

"Thanks for allowing me to take a little of your time—"

"I tell you, I don't know why you're doing this, young man, since Channel Seven has my story." She suddenly lit up, her stern expression changing into an animated one. "But since you're here, how about a glass of Cold Goose, before we get started?"

Against his better judgment, he nodded. Based on her breath, she had already started, and much earlier. Not accustomed to drinking at this hour, he rationalized she'd open up more if he cooperated with her. In his line of work, if you were standoffish, you were professionally dead.

She returned a few minutes later, carrying a plastic tray emblazoned with the logo of the Sesquicentennial celebration of the city's founding, with two fluted champagne glasses on it.

He took one of the glasses, the one with a chipped rim, and sat down.

"I'm seventy-eight years old and lived my whole life in Harrisburg. This has never happened to me before," she said, putting the tray down on the coffee table and picking up a glass.

"All the more reason I'd like to hear your story."

"I arrived at the church basement, you know, St. Patrick's Cathedral Chapel on State and sure enough, they have these new-fangled voting machines," she took a generous draught of the sparkling wine. "This is the second time I voted with those machines, and the second time I've had a problem."

"And what is that problem?" Paul asked, before taking a sip of the effervescent liquid. It tasted insipidly sweet, like the fruit wines he and his buddies drank during their first hoedown. But tepid, with bubbles. Yuck.

"Last time, my card didn't work. This time, well, let me tell you what happened. I voted for Duquesne in this election." She turned to him. "But I suppose you can't tell anyone that."

"That's right, I will not tell anyone, unless you want me to, Mrs. Muller."

"Well, if you must. Oh, and do call me Marie," she continued.

He once again detected the smell of alcohol 'dehydrogenate' on her breath. According to what Elaine had once explained to him, her liver was, slowly but surely, decaying.

"So when the summary page came up, I had a second thought. I saw 'Dusquesne' highlighted. I decided I didn't like that Duquesne fellow after all. So, I pushed the button marked – 'Modify' next to the box marked 'Duquesne'." Paul took a sip. "Well, the sheet went blank and then came up with all the names, just as I had first voted, without any 'X's' next to the Senate candidates. So, I touched that box on the screen next to 'Dooley,' and an 'X' appeared next to him. Afterwards, the same summary page came up." Muller picked up her glass and downed half of it.

"You mean with Duquesne's name?"

"No, no, with Dooley's name this time. But then I thought to myself, 'I don't like the Dooley children. They're so . . . proud.' Did you see them when they were asked about what they would do to make Harrisburg better?"

"No, I can't say I have," he said, suppressing his desire to fidget.

"All they could talk about was those parks for new fangled skateboards and lowering tax rates, cutting benefits for people like me. So, I pushed the square that said 'correct'. No, no 'modify', that's it. The new screen came up and so, again, I pushed the square next to 'Duquesne.'"

Paul glanced at the magazines on the table. He didn't recognize any of the covers from newsstands.

"Well, 'Review all votes' appeared on the right side of the screen," Muller continued, before pausing, apparently losing her train-of-thought.

Paul turned his head slightly to the right and squinted with his right eye in an attempt to see the dates on at least one of the magazines. *Yes.* The closest magazine to him was over three years old.

Muller regained her thoughts. "So, that's right, the summary page showed an 'X' next to 'Duquesne.' I carefully reviewed every vote. You know, I'm very thorough."

"No doubt." Paul nodded.

"But when I touched the box that said 'press here to send vote,' the screen went black for an instant. Then another summary page flashed on the screen for two seconds. I saw only the first line and saw an X next to 'Dooley'. And that changed vote was sent . . . to the machine's brain, I'm sure of it!" she concluded, downing the rest of the wine in her glass.

"Some more?"

"I'm fine, Marie, thanks."

"So I voted for Duquesne, but that, that *machine* changed my vote to Dooley," she said, gesticulating with her finger, her head shaking. "You undershtand, I had no choice but to go, to go to the election judge. So, a nice black lady—you know, they're not usually nice but thish one was – took me to him."

Paul nodded. He had to cut to the chase before Muller passed out. He leaned in to her. "Marie, as a reporter, I have to ask this question. You're quite sure you didn't change your vote by accident?"

Muller looked at him, he surmised, debating whether or not she should claim offense and insult. She vied, he was pleased to see, to soldier on without drama queen episodes, though her facial expression lost its playful character.

"No. I changed my vote from Duquesne to Dooley and then back to Duquesne, because he has a nice wife and I like hizh children more than candidate Dooley's, but the machine changed my vote back to Dooley."

"And your interview?"

"Well, some bigwig pulled up in a, a limousine and took the reporter and—" she hiccuped, "me to the sht-udio."

"Studio. Uh-hunh, and what did the studio look like?"

"There was a camera with lots of, of cables and three men around the table," she said, awkwardly lifting three fingers. "Oh and two of them were the handshomest I think I've . . . ever seen," she added, coquettishly smiling, her eyes oddly unfocussed.

Ten minutes later, he bid Marie farewell and stumbled down the steps. The buzz helped him adjust to the cold.

As he drove away, he remembered a favor he had done for the associate director, covering some birthday party of a mucky muck on a Sunday. The story was vapid, but since it had involved log-rolling and traditional greased-pole climbing, it had just enough meat on it to be reportable. He needed to collect on that favor, before he became a distant memory to WQHP, which, he figured, was about five days on the outside. He pulled out his phone.

"Hi, Pete? This is Paul . . . "

"Yeah. I heard the news. Sorry, man."

"It happens. Hey, how's about them Penguins this season?"

"Great, Paul, but – uh – I'm kinda busy now."

"Sure Pete, I was just wondering, could you find out who from Channel Seven interviewed a lady last night about a problem with her machine? Her name is Marie Muller."

"The one you missed?"

"'Missed' is quite a loose interpretation, but yeah, that one. Oh, and the feeds," Paul asked, crossing his fingers.

"You can forget the feeds. You know how things work. But I'll try to find out who covered the story."

"Super."

"And Robinson, after this, we're even, okay?"

"Sure, sure, Pete. You're a darling."

7

Jenkins was still dozing, and didn't bother getting up when he felt his administrative assistant lifting her head off his outstretched arm. He'd just re-entered the best part of his dream, where he and his buddy were getting a full belly-dance routine from a couple of cute Syrian chicks. He was holding a beer in one hand, wads of tens in the other. In the background, over the fence, stood the Buratha Mosque and his captain was bawling out a call to prayer.

He heard a door opening behind him in his dream and jerked his head up.

His admin assistant turned her head towards him, "I, um, gotta go to work. Had a nice, uh, time."

Jenkins yawned, and rolled to his side, resting his head on his right arm. "Yeah, me too, Jenny . . . See ya around."

She hesitated pulling the door, half-smiled and left.

***

An hour after interviewing Muller, Paul left for the airport. He still had three days to wrap up any business. The station would need to assign other reporters to the human interest beat, and he would have to hand off the stories-in-progress. By now, WQHP was, thanks to his pioneering approach, the go-to channel for small feel-good stories. 'The Nijinsky of bubble-blowing stories' he'd once called himself. But that comparison he had made under different conditions, and, he had to admit, that's not really why people watched the news.

So far this afternoon, he had drunk only coffee, chewed only Advil and swallowed only his pride. He collapsed onto the leather couch in the lounge. Before he could fall fast asleep, his mobile phone rang. He fumbled a few seconds, before clapping it to his ear.

"Yes. I'm listening," he said, yawning. "No, no, a no-martini lunch. Whatcha got, my dear Watson?"

"Wise ass. You're lucky the assistant producer is a pal. He looked into it and no such interview. He'll check with one last reporter, but yeah, that's it. No Muller, no feeds, nothing." The associate producer paused, "Like there wasn't a story there at all. In fact, he was pissed at me for wasting his time . . . and I have only you to thank for it."

"Thanks, Pete, you're a gentleman and a scholar. We're now officially even."

"Yep. And Paul, I hate to break this to you, but you can't access research anymore either."

"Me, access research? Wouldn't think of it. Luv ya, Pete. Bye."

As he approached the gleaming Danabold Technologies building in the rental car, Paul saw no free spots in their parking lot. The strip mall on the other side of the street looked promising, so he turned into that parking lot and kept driving towards the back of the lot until he found a spot next to a Fuddruckers.

The receptionist in the burnished metal-with-teakwood-California-inspired office was wearing what he surmised to be a Gucci jacket and could have been a fashion model.

She looked up at him as if he were the rodent removal specialist from Rentokil. "Hello. Are you the reporter?"

"Right you are. Paul Robinson."

"Mr. Harrison will be with you shortly. Would you like a coffee, espresso?"

He now felt surer that no one would recognize him here in Ohio. They might have seen his story on the Olympic swimmers, but nothing else. For now that was good. He needed to fly under the radar until the story had some foundation – if that even existed. "No thanks, any more caffeine and I'll reinvent the internet." He paused for a reaction.

None. Same expressionless face.

"Maybe a water," he added.

"Sparkling? Still?"

"Still."

She clicked a button on the side of her headphone. "Roger, could you bring in an Everclair and a Chai, please."

A well-tanned man with longish hair, carefully moistened and tied back neatly, walked in, smiling infectiously. He was dressed in an expensive European-cut suit. "Robinson, WQHP, I presume?"

"Right you are."

The man stepped around the receptionist's desk and approached Paul, smiling as if he had found a long-lost friend. "So pleased you could come visit us." He looked behind Paul. "What, no cameras?"

"No, for now I'm just researching if there's a story here or not."

"Fine, come on in."

Paul followed the smooth-sauntering man into an enormous office, lined left to right with awards and framed full-color press releases. Harrison waved him to a leather armchair, before sitting down behind the polished cherry wood desk.

"Mr. Robinson, it is truly an honor . . . to receive you here," he said, beaming a smile and smoothing back his gelled hair with his right hand. "But before we get started with your questions, I have one for you."

Paul gestured to go ahead.

"Do you know how Danabold got started?"

He knew that saying 'no' would bring on a litany of useless information, but aside from its being a small subdivision of a multi-billion-dollar enterprise, he really didn't know anything about the company. His pause apparently served as a confession, and he was soon learning about the company's origins in Dallas, Texas.

It had been started by two re-plants from Southern California; people who saw great promise in modernizing processes that were still carried out the nineteenth century way. So these bold men borrowed technologies arising from the development of ATM machines, EZ Pass scanners and even gas pumps, to develop Direct Recording Electronic voting machines. "Of course, Danabold hopes that our voting machines become the standard, and that electronic voting, or more precisely D-R-E voting, will eventually be adopted by enough states for the U.S. Congress to finally pass a law standardizing the actual mechanics of the voting process."

42

"Naturally," Paul replied, biting his tongue.

"As you may well know, today any district can employ whatever electronic or paper method their delegates choose. It can be hand-written ballots, or optical scanning of . . . hand-written ballots, now how dumb is that? Electronically scanning error-filled forms," Harrison said, rolling his eyes. "Or they can use our state-of-the-art machines, simple and straightforward, voting by touch. Total vote accurately counted and reported within thirty seconds of the close of the polls. Can you believe they're even debating it?"

"Indeed, it is unbelievable, Mr. Harrison. But I have a question. How did you test these for ease of use?"

"Aha, a man who cuts to the chase. I like that. We tested with over one thousand people, a cross-section of the population, in real-time tests, before the actual election."

Paul proceeded to ask half-a-dozen questions about the cross-sections of population, which should reflect the population, as well as testing methodology.

The responses were, like the furniture in the office, polished, well-crafted.

"And this cross-section included how many retired persons again, as a percentage?"

Harrison cocked his head slightly, as his lips tightened for a second. He smiled broadly. "You asked the right question. This demographic is by far the slowest to adopt technical innovation. A third of the test populace was over sixty, more than actually vote. Of course, prior to that, during development, we consulted separate focus groups of senior citizens. As a result of our findings, we made some modifications and re-tested," Harrison shifted his weight, making the polished leather squeak in an irritating manner. "For example, our screens are twenty percent larger than the Harker equivalent. Ninety-three point-nine percent successful after all those changes. Now what odds-maker wouldn't want those odds?"

Paul nodded and looked deep into Harrison's green eyes. "Certainly none that *I* know!" he retorted with the eagerness of a ten-year-old. "Now, Mr. Harrison, I'd appreciate it if you could send me those details on the testing with seniors."

"Certainly. Anything else?"

"I have just one more question."

"Shoot."

"How can you be sure that your machines won't be tampered with? I mean, you heard about the Sequoia machines in Arizona being hacked, right?"

Harrison's eyes narrowed. His neck muscles began tightening.

"You know, Danabold will not comment on the weaknesses of competitors' machines. There are lots of companies that think they can program secure voting machines. And it just gives the industry a bad name, as well as hampering the inevitable revolution in voting technology. It's advancing with kicks and starts as is."

"Uh-huh. So, Mr. Harrison, there's no way someone could actually verify if a machine had been tampered with."

"Of course not." Harrison smiled widely, shaking his head. "The machines are sealed."

"But the seals are verified?"

Harrison nodded.

"What about the possibility of hacking from the outside?"

"None at all. Nada, non, nyet!"

"Mr. Harrison, let me rephrase the question—would you be willing to be quoted on national television, confirming that your Danabold Fast Pass model is one-hundred percent hack-proof? Nobody can flip votes with it."

"Absolutely, a hundred percent. With a triple key card design, it would take someone to line up all three cards, and that is physically impossible, since the third card rests with the CEO of the company."

"So that person would have to be physically present to change the programs once they are set."

"Absolutely."

He knew enough about the internals of corporations and boards to know that the CEO wouldn't hold onto anything—even the Coke recipe was kept in a vault. As far as Paul was concerned, this interview had been over fifteen minutes ago. He nodded, silently.

"Well I happen to have some background information here that you might want to consult . . . should you have any more questions," Harrison said, handing a sturdy ring binder, crammed four inches thick with multi-colored paper, to Paul.

"Well, thank you, Mr. Harrison. Something to build my biceps with," he said, smiling.

The PR man looked blankly ahead.

Paul smiled even more widely. "I don't think there's anything but a great story here. Couldn't really add to what Business Week did last month. I'm sure you're busy—"

"Meetings with reporters of your stature, Paul, are my 'busyness,' Like oxygen to a scuba diver."

"Oh, Mr. Harrison." Paul guffawed theatrically. "How could you end on a better note?"

"By you wanting to see the machine?"

"Thanks, but I already have, when you demonstrated them three years ago at the Hilton in Pittsburgh. I have to say, I was impressed."

Paul smiled and squeezed Harrison's hand tightly, to ensure he communicated sincerity.

He was happy to have left after less than forty-five minutes. *Oxygen to a diver . . . good one.* Paul had no story. The company had tested with retirees. Muller was, he had to admit, a lush. On the way back to the car, he dumped the binder into a trashcan. *PR departments are a scourge.*

He looked at his watch. 4:50. He had some time. At the very least he'd have a burger and catch part of yesterday's hockey match . . . if he couldn't catch more.

He strolled into the near-empty Fuddrucker's establishment and sat down at the main bar at the internal left corner where, he surmised, he'd be a wallflower. He ordered a Samuel Adams and settled in, watching the Blue Jackets battle it out with the Red Wings on one of the three screens above the bar.

Ten minutes later, customers started streaming in, filling tables, and some sat at the bar. Two guys in their late twenties came and sat at the other end of the bar, facing another screen.

"Hi, what can I get you fellas?" the bartender asked.

"Hey, Chuck. Anything to get the taste of 'same-old-shit' out of our mouths. A Fat Head should be good for a start," answered the shorter of the two.

"Make that two," the taller one said, staring at the screen. "Oh yeah! Last night's game."

"Dja catch it yesterday?" his friend asked.

"Nope, not a chance. We submitted the 'B of A' ATM proposal today."

As dozens of other workers descended upon the establishment, an idea popped into Paul's head. He got up and sat one stool away from the shorter of the two guys. He stared intently at the screen, along with the other two.

"Did you see Nash make that shot off Korda's shoulder last week?" Paul piped up.

"Only about fifty times," replied the shorter one, still staring at the screen.

After five more seconds, Paul sensed the guy was staring at him. "Hey, haven't I seen you before?"

"Quite possibly."

"Aren't you that reporter, that one on YouTube, the Pennsylvania swimmers' story?"

"Yep . . . when I'm not a professional Penguins fan."

"Penguins?" the taller one asked.

"Yeah. I've scraped the yellow paint off my cheeks so many times, it feels like I'm scraping my scrotum with cat claws now every time I shave."

The short guy laughed outright while the tall one smiled.

Holds things tight to the chest, the tall one.

"Name's Paul Robinson, WQHP Harrisburg."

"Chris," said the short guy, "and this is Jim. So, what was your funniest story?"

He unfurled his all-time favorite reporter-on-the-beat escapade. After a minute the two newly-made friends were leaning forward, oblivious to the game going on behind them.

"So the chimp was going to hug me and then sit on my lap, right, while I told the story about the zoo needing funds."

"Did they sedate the chimp first?" the tall one interceded.

46

"Hell, no! And since this was my first ever chimp encounter, I didn't know shit," Paul conceded. "So, of course, after he hugs me and sits down, I start feeling this hot wet liquid on my pants . . . and that was just the start. I mean he starts pissing on my lap with this turbo jet stream! Swear to god, I still have a welt to prove it. And all this time, the trainer is calmly nodding, as if *nothing* is happening, or worse, like *I've* got this bladder problem he's gotta ignore, you know, to be polite."

Chris doubled over, laughing, while Jim smiled.

After Chris' laughter subsided, Jim cleared his throat. "So what are you covering now?"

"Nothing at the moment. But sitting here I had an idea, aside from ordering The Works burger —"

"That idea's a good one, you won't regret it," Chris interjected.

"Lucky guess . . . So my other idea, is to do a series on how e-technologies affect politics. You guys aren't from Danabold, are you?"

The two looked at each other.

"Who isn't, in here," Chris said.

"You know anything about those electronic voting machines?"

Chris jerked forward as he scraped the bar rail with the bottom of his feet. "Naw, not really . . . most of that's classified, high-security process stuff. Info should be available from the PR department."

"Sure, I just talked to this guy named Harrison, but," Paul leaned in, "I need to supplement that with a real story from real programmers, not just a talking PR head's press release. I didn't rely on just the state's Olympic Committee for the swimmers' story, if you know what I mean."

The two guys exchanged nervous glances. Jim furrowed his brow.

Paul leaned well into Chris' comfort zone. "To tell the truth, guys, it means my job to me. The boss-man said if I don't deliver on this story, I won't be drinking eggnog at the Christmas Party," he stated in a low, confidential tone, "—'cuz I ain't gonna be around for it."

47

Chris looked up at the TV, as if avoiding the gaze of the bag man who has just thrust out his palm for some change.

"Wish I could help," Jim finally said, nonchalantly.

"Come on," Chris countered, "let's give the guy a break. He's gonna get fired. Besides, he flew all the way out here for it."

Jim stared at the screen for a few seconds, before turning to Chris and nodding.

"Okay, but we didn't tell you," Chris started, still facing the screen.

"I know how to protect my sources."

"One guy's name is Leboeuf. Don't know his first name. Murray or Marty or something with an 'M'. Apparently a real square peg. Then there's Kendrey," Jim said, in a low voice. "Totally bonkers, but fucking smart."

"Yeah, has some mental problem; schizo or something," Chris interjected.

"Kendrey – is that a first or last name?"

"Josh Kendrey," clarified Jim, before adding, "rumor has it, they're working with guys from other e-voting machine companies on some big project in your state, but we don't know jack. Anyway, the two of them moved a few months ago to Harrisburg. Closer to the action."

"Know where they're staying?"

"No, but I think someone mentioned Leboeuf is from there originally," Jim answered.

Jim turned to Chris. "Travel would know where Kendrey is staying, wouldn't they? Do they work this late?"

Chris nodded. "Yeah, what with board members flying off somewhere all the time." He pulled out his mobile and pushed a button. He turned back to Paul. "It'll cost you a beer, reporter-man."

"Got it."

"Hi, Margie, this is Chris . . . Yeah, the flight to Baltimore was fine. Glad I'm back, though... No, I didn't try the crabs. I was on a clam chowder budget. Listen, Margie, I heard I might be going to Harrisburg, Pennsylvania. Where would I be staying? I wanna make sure the place has a gym. Yeah, thanks, I'll hold."

48

Chris nodded to Paul, and put up his finger. "It ain't the Four Seasons, tell you that much." He glanced up at the screen as the Red Wings' forward drove in a goal. He winced. "One of the apartments at the Executive Towers? Uh-hunh. Uh-hunh. Do they have a gym? Rats . . . oh well . . . I see, one across the street. Thanks a million, Margie . . . Yeah, if I go, I'll bring you back a pretzel."

He turned to Paul. "Some long-term arrangement with the Executive Towers. You know the place?"

"Sure. Makes sense, it'll save the company money if some guy's going to be there for a while."

"Saving money is what this company's all about," concluded Jim, looking up at the screen.

The three went silent as the forwards of the Blue Jackets maneuvered the puck to within 40 feet of the goal. A ricochet goal from thirty feet caused an uproar in the bar, breaking the silent pause.

Paul finished his burger and looked at his watch. He got up off his seat and took out his wallet. "Thanks a mil, guys. Flight's leaving at half-past six, so I gotta run. The drinks are on me," he said, putting thirty dollars on the bar top. "And next time you're in Harrisburg, look me up . . . I'll treat you to some real beer!"

"Yeah, so you're telling us you've got the best hockey team and now the best beer? Bullshit." Chris blurted out. "Just make sure to tell 'em that Danabold programmers rock!" He smiled.

"Rock? Hell, they're the Clooney's of the industrial world!"

The plane landed just before eight. Paul wondered if it was too late to pay a visit to the programmer. He decided it wasn't. If the company had reassigned the guy to be somewhere else right after the election, the guy could be gone by tomorrow morning.

Paul parked in front of the Executive Towers apartment building. The building, built in the '80s and partially renovated since, still afforded nice views of the river from the third floor up, a major selling point of a structure past its prime.

As he shut the car door, a flowerpot exploded on impact, ten feet behind him. He felt pottery shards hitting his pants. He swung around and looked up. *Come on, throw down another one, asshole, and let me see you.* Paul clenched his hands into fists.

49

No open windows, and no one visible.

He kept looking upward as he returned under the portico. Once again he approached the intercoms. He read the names. Two of the forty-odd intercom buttons had 'Danabold' written next to them, and the one between them was blank. He guessed that Kendrey, as a programmer with a crazy streak, wouldn't be the first one. He pressed the bottom one, and waited. There was no answer. After ten seconds, he pressed the top button of the three.

"Yes?" someone answered.

"Paul Robinson. WQHP news. Is this Joshua Kendrey?"

His question was followed by silence.

"Hold on," the man answered, before hanging up.

Paul waited. The wind was swaying the trees. He decided to push the middle button.

After a few seconds, the intercom went on. Led Zeppelin played loudly in the background. "*Yeah?!*"

"Joshua Kendrey? It's Paul Robinson, reporter from WQHP. Would you have a moment to talk?"

"Do you know I just burned my pancakes? I hate you, I hate all of you!!"

"Look, I understand, I hate some reporters too, but I'm not like them, trust me." And then Paul delivered his fluff bomb, the one that never failed to disarm his prey: "I'm a human interest reporter." Paul heard no answer. "We're putting a special together on the brilliant folk who program computers for cutting edge apps."

"Yeah, I'm brilliant awright, bu-bu-but why would I wanna talk to you? It's not a good time r-right now for me."

Paul thought he heard a sob, but it was hard to tell through the intercom. "I'm a pathetic low-life, whose every motion is dictated by little fucking yellow and green pills!"

"Would you wanna talk about it?" Paul asked as gently as he could.

"Sure, on the condition that you bring back my flowerpot *in one piece!*" the shrill voice responded.

"Look, throwing flowerpots from five stories up is dangerous. Didn't your mom ever tell you that?"

"Don't bring my mother into it. I missed you by a mile anyway."

50

"Okay, okay. Look, Josh, if right now is a bad time, I'll go talk with Leboeuf. He seems to be the brain behind the programming anyway."

"Marty? Brain? He couldn't program Pong. Good one, anchorhead . . . Oh no, the spirit comes. Aargh. Leave me alone!" The click was audible as Kendrey hung the intercom phone up.

Paul shook his head and slammed the wall with his open palm. The reverberations set off a painful reaction midway up his forearm. He rubbed his arm.

A well-tanned man walked across a lounge furnished with beige, stained sofas, and to the glass door. He opened it, but didn't invite Paul in.

"I'm of Danabold, and you are?" he asked.

"Paul Robinson, WQHP news. I'd like to speak with Mr. Kendrey."

The well-tanned man stared coldly at him. "How did you hear about Mr. Kendrey?" he asked.

"I talked to someone at the Department of State, someone who had ordered the D-R-E machines," he lied, attempting to disarm the man.

"And what would you like to ask him? He's very sensitive to outsiders, and, like many truly gifted people, he has a short fuse. Think 'Rainman' with Dustin Hoffman," the man said, tightening his tie.

"Would you know—"

"I'd recommend you speak to our Public Relations Director," the man interrupted, "he has all the information you'd want, and more. Mr. Harrison is his name."

"Just spoke to him a couple of hours ago . . . good back story."

The tanned man pursed his lips and folded his arms over his chest.

"Look, mister, it's no big deal, I just wanted to get more info on the machines Pennsylvania was using in ten of the districts. And more importantly, thought I'd do a profile on the minds behind the programs. All too often, people forget that a human being programmed the machine, a human with hobbies and concerns."

"Yes, yes, a worthy story, Mr. Robinson. Why don't we do the following: I'll ask Mr. Harrison to arrange a meeting for you with the team of programmers who made this machine in Ohio. Same machine, same insides. But they'll talk to you in a language you can understand. With Kendrey--" he started, shaking his head, " . . . it's all quants and binaries and, uh . . . insults, if the truth be told."

"Sounds great. I'd be happy to talk with the team," Paul countered, against his will.

"See, we can always find a way to accommodate the press. Could I have your contact number?"

Paul pulled out a card from the stack that had only his office phone number, and passed it to the man.

The bronzed man handed his over, and he glanced at it. 'Jacob Lewis, Senior Director of electronic voting and tabulation machines, Mid-Atlantic States.'

A tenant approached the door and stared at Lewis, apparently impressed with the late-Fall dermal patina. Lewis smiled a good-bye to Paul and held the door for the tenant.

The tenant glanced at Paul. He did a double-take.

"Hey, aren't you Paul Robinson? You did the Old Adams County polling station story, right?" he asked.

"Yeah."

"Bo-ring. You should stick to chimps."

8

The last of the five websites Paul accessed on his computer was the Mayo Clinic site. He realized he hadn't a clue as to whether Kendrey would be classified bipolar I or bipolar II. "Bi-po-lar," he let the phrase run off his tongue. *Sounds like a white bear with a sexual identity issue.*

In fact, according to the symptoms description, he'd been 'bipolar' himself at least a half-dozen times in his life. He picked up his phone and speed-dialed Latanya's number.

"Hello?" she answered cautiously.

"Tan, it's me."

"Yeah, I know," she replied briskly, "what do you want?"

"Another chance."

"You know, Paul, I'm reading this awesome book, "*It's Called a Break-up Cause it's Broken.*"

"It couldn't be about us."

"Well, maybe you should *investigate* it, hmm? Rule number one, no contact for thirty days."

"Is that what you want to do?"

There was a long pause on the other end of the line.

"I dunno."

"I'll get the story. Stations will be beating a path to my door."

"Honey, I've heard that song before. Listen, when can I pick up my stuff?"

"Tan —"

"When will you not be around?"

"Okay. I'll be out all day tomorrow."

He hung up his phone. He tried not to succumb to the impending feeling of loneliness. He stared at the screen for a full half-minute before speed-dialing another number.. "Elaine, hi, it's Paul."

"I got the news. I'm sorry to hear it. The station certainly won't find anyone as good as you for those stories," Elaine interjected.

"I dunno . . . Maybe I'm losing my mojo, but gotta keep keepin' at it, I guess. Thanks for the support anyway. But I was actually calling you about something else."

"Shoot."

"Remember that social worker friend of yours, the one who worked with seriously depressed people?"

"Paul, don't play drama queen. I know you better than that," Elaine added.

"No, not for me. It's for this story I'm investigating."

"Fine, whatever. His name is Lee, Lee Durand. He works at the Rose Institute. Usually the graveyard shift. Just a second, I'll give you his number."

Paul woke up at quarter past five. He scratched his scalp and yawned loudly. He got up five minutes later and padded to the kitchen, strolled past the 'Physician's Desk Reference' on the counter and to the refrigerator. He drank the orange juice straight out of the quart container, and wiped his lips with his forearm.

"Well, Kendrey," he said out loud, "congratulations, you're more fucked up than a football bat." He picked up the cell phone and dialed Durand's number.

He wasn't at the first clinic, but the nurse gave him the number of the second one.

Paul dialed the number the nurse gave him and looked at the clock. It showed 5:45. "Hello, may I please speak to Lee Durand?"

"Yeah, speaking."

"Hello, Lee. This is Paul Robinson, from WQHP and I need your help."

"Yeah, I'm not surprised. How can you even sleep at night for all the dirt you dig up on people?" the man let out in staccato. "First thing you should know—I don't do recreational hallucinogens."

"Wait, wait, I'm the human interest guy. Besides, it's not about me. I need to work with, to help, a guy with bipolar disorder."

Durand agreed to see Kendrey at nine in the morning. He'd carry out an assessment for sixty dollars.

As a journalist, Paul had been taught to use open-time windows, even half-hour slots, efficiently. He picked up the phonebook and found three Leboeufs listed in Harrisburg. The one on Tenth Street, not a chance—bad side of town. A second one lived not far from his neighborhood, but the tonier sub-section near Kohl Memorial Park. Jameson Leboeuf. *Maybe, but that would mean he came from money.*

Third Leboeuf, in Cumberland. The right place for an ambitious young corporate guy, especially if he had a family. The first name was . . . 'Martin.'

It was seven-forty in the morning. A little early, but not unreasonable and, more importantly, would catch the programmer before he left for work. Paul stood on the front porch of a handsome four-bedroom house, noting to himself how quaint pink flamingos looked in the right setting, the present one not figuring among those. He patted down his hair as he waited; he didn't remember brushing the diminishing strands that morning.

Paul noticed someone's eye covering the peephole, as the dot in the middle changed from white to black. The light dot reappeared. After another ten seconds, the dot went dark, and he heard a wavering voice

"Who is it?"

"Paul Robinson, WQHP News."

"What do you want?"

"We're doing a story on hi-tech programmers and the innovations that are changing--"

"Not interested. Go away," the voice responded gruffly.

"But this is a great chance to inform . . . "

"NO!"

"Your voting machine—"

"Get lost!"

Paul put his ear to the door. He overheard a female voice saying something about a reporter and Olympics and swimmers. This was followed by an object being thrown, and then a resolute "No. Don't let him in!" Then a "Honey, what is your problem?"

Paul turned on his heels, when he heard one door bolt open, then another and the door opened. Mrs. Leboeuf looked the good suburban housewife and rotund caretaker of the progeny, with a bright yellow bow in her neatly-coiffed hair. "Come on in," she said merrily.

Paul entered. "Hi, I hope I'm not too early."

"Not-at-all. It's a pleasure to meet you . . . I loved your story on the Olympic swimmers. Have you had breakfast? Let me just bring out some coffee and Danish. Marty will be right out," Mrs. Leboeuf said in a sing-song voice that defied a negative answer.

"Sure, coffee sounds good, thanks."

Mrs. Leboeuf quickly hurried back to what he assumed was the kitchen, on the other side of the wall to his left. He heard plates, cups and saucers being brought down and arranged on a table. However, Martin was nowhere to be seen.

Paul looked around the living room. A souvenir from the Penn Caves graced the right-hand wall; some Hummel statuettes on the fireplace mantel; a big plastic car in the middle of the room, together with half-a-dozen soft animals strewn near and far. *Kids, the beginning of the end . . . but maybe it's time.*

After another awkward thirty seconds, and seeing no one, Paul stated loudly: "Well, Martin, you needn't be modest. What you did at Danabold—"

Leboeuf emerged from behind the back wall, a short-nosed revolver in hand, pointing it at Paul's forehead. He cocked the pistol.

"Get the Hellouttahere!"

He had never stared down the barrel of a gun before. Not from four feet away, at least. He felt the blood drain out of his head, his arms breaking out in goose bumps.

"Okay, okay, I will . . . just be cool." He slowly lifted his right arm. Without taking his eye off Leboeuf, and keeping his right hand up, he retreated two steps, fumbling for the door knob with his left hand. He turned the knob slowly, all the time focusing on Leboeuf's right eye.

Mrs. Leboeuf ran into the living room.

Paul ignored her, opened the door and stepped outside. The last thing he saw before shutting the door was her terrified face.

Paul jumped in the car and pulled away from the curb, wheels spinning. He drove three hundred yards down the street before pulling over again. He lifted his hands from the steering wheel.

They shook like the last of the orange leaves clinging to the maple tree in front of him.

9

"The reporter went to see Leboeuf this morning. Last night, he tried to get into the Executive Towers. He goddamn knows something . . . I just hope not about Duprix."

Jenkins started thinking from the other side of the equation. *Had this reporter been to the Department of State?*

"Well, now," Jenkins started in a relaxing banter, "however the cat's done stuck his head out of the bag, we gotta cut that head off now, don't we?" Before leaving, he wondered if this guy would be expendable, too. He concluded that he must be.

***

On the street outside the Leboeuf's house, Antoine got out of his Chrysler Crossfire. He wore calfskin leather gloves and he appreciated their softness. He walked to the door and rang the bell. He fingered the Beretta PX4 Storm in his coat pocket. The silencer was short and fat. With that attachment, the gun didn't stick out of his pocket.

"Who is it?" answered Mrs. Leboeuf.

"A colleague . . . from Danabold Technologies."

"Please wait a . . . minute," she said. He's occupied now. Perhaps you could—"

"It's his caretaker."

"Caretaker?"

"Yeah, ma'am, could you just ask Marty to come to the door." Antoine waited half a minute. He then heard someone scratching the door on the inside.

"Look, Mister. I-didn't-say-a-thing! This—" Leboeuf sputtered out.

"Relax, slow down, Marty . . . Could I come in, it's freezin' out here."

"Sure, sure, sorry."

The door opened. Antoine walked in slowly. He took his hands out of his pockets, glancing towards the kitchen and the stairway and then stopped on the edge of the Karastan carpet that covered the living room floor. He stood with his legs slightly apart, and his arms loosely at his side.

"You were sayin'?"

"Some reporter came from WQHP—but I didn't say a thing. Not a thing. He started asking questions, so . . . , so I took my gun, and . . ."

"And?"

"I took it out and pointed it at him. He left in, in a hurry."

"Did you fire it?"

"No."

"What about the police?"

"Are you kidding? I saw what happened to Duprix."

Antoine nodded.

A child cried out in the kitchen. Two seconds later, the two-year-old boy ran into the room, took a look at Antoine and then walked slowly to Leboeuf, keeping his eyes on Antoine. He hugged his father's right leg. Leboeuf stood, frozen. *He wants to pick up his son. Why doesn't he?* Antoine looked at the boy and smiled.

The boy continued looking inquisitively at him. Antoine knelt down on one knee and patted the child's head. He slowly reached into his pocket with his right hand.

Out of the corner of his eye he saw Leboeuf stealthily moving his right arm behind himself, his cheek muscles twitching. Keeping his eyes focused on the child, Antoine slowly pulled out a bright red shaker toy, shaped like moose. He held it up at eye level to the child, so Leboeuf could see.

The boy smiled, grabbed the toy and ran back to the kitchen.

"Okay, Marty, I'll believe you . . . this time. I know you got a lot riding on it," Antoine said, slowly getting up.

Leboeuf's eyes widened. His face was pale.

Antoine smiled wryly, took a step towards Martin and stopped with his lips below Martin's ear. "Just a word of advice, kid," he stated calmly, "if you're not more careful with that gun, you might find yourself with a hole in your head one a' these days." He waited until Martin dropped his gaze. "Got any problems, best you call you-know-who and ask for the 'caretaker.' That's me."

59

He waited until Martin nodded. "Good. Now I got instructions for you."

Leboeuf took the small envelope from Antoine.

Antoine turned on his heels and stepped towards the front door. Then, turning his head towards the kitchen, Antoine bellowed: "The Lord is your shepherd, ma'am. Maybe next time, we can meet face-to-face."

As soon as Antoine got into his car, he punched his cell phone keys.

"Good news, chief."

"What did Leboeuf tell him?"

"Nothin'."

"You believe him?"

"Yeah. He knows about Duprix. Scared to death for his family. Even pulled heat on the reporter."

"Okay. What should we do?"

Again, the pinnacle a' doubt and indecisiveness. How could he be anything but a tool in the organization?

"Nothing. He's gonna do what you ask, and he ain't gonna squeal."

***

Paul pushed the intercom button while Durand took out a small spiral notebook.

"Josh, it's Robinson, the reporter."

"I'm not well. Go away," Kendrey replied.

"Wait. Joshua, I'm from the hospital. I think I can stop the voices in your head," Durand said.

"Okay, but be fast. They're gnawing through m-my cortex!"

The door buzzed open. The two walked in and continued across the lobby carpet, past the vase containing artificial flowers on the faux-gilded center table, to the elevators.

Paul steeled himself to receive a wave of body odors, rotting Chinese food and stale pizza. But when the door opened, he glanced past the tousle-haired man to see a spotlessly clean apartment.

"You must be the reporter," said the roughly forty-year old man, wearing a clean polo shirt and jeans. "Do you p-p-play Scrabble?" he asked, as he led them to the living room.

At a table stood five computer screens with Scrabble boards on them; the games were in varying stages of progress.

Paul looked at Durand, who just shrugged his shoulders.

"Occasionally," answered Paul, still off-balance from the welcome. "Listen, Joshua, could I ask you some questions about your last project at Danabold?"

"The gremlins won. They were backed by Nazi infiltrators . . . If you want to play, go to board three. The game just started, and the Hun has a cold," Kendrey responded.

"All right, board three. Now, about the voting machine project," he persevered.

"No. The enforcer told me to talk with no one about it. Zip. My lips are sealed."

"Who's the enforcer?" Paul asked.

"Internal security."

"Could you at least tell me if it would be possible that the machine inadvertently changed votes, or prevented some from being changed?" Paul waited a full twenty seconds, before Kendrey cocked his head to the left side and stared at him.

"No way."

"Not according to at least one lady here in Harrisburg."

Kendrey rolled his eyes. "Some old bag? Betty White wouldn't have had any problems."

"The Republican victory was larger than polls had predicted, especially the U.S. Senator's seat."

"Aargh. You stink. I don't care about politics. I program computers!"

"Okay, okay, would you like to talk with Nurse Practitioner Durand?" Paul offered.

"Sure," Kendrey responded. "Nurse, do your pants ever feel like they're on fire? I'm so tired, I've gotta rest. I've got a busy afternoon ahead of me . . . the championship, you know."

Kendrey suddenly stiffened up. He was perspiring. Without another word, he trotted off to his bedroom.

61

Durand followed and Paul filed behind. In passing the kitchen, Paul noticed that the counters were equally spotless and unburdened by open boxes or jars.

"Joshua, what medications are you taking?" Durand asked.

"Xanax, Citalopram, lithium."

"When did you last take the lithium?" Durand continued prompting.

"When Elvis died."

"Joshua . . . " Durand interrupted, sounding annoyed.

"Okay. Maybe a little less. His body was in late stages of decomposition. Worms were coming through his eye sockets."

"Joshua! . . . When?!"

"This morning . . . every morning, at eight, for the last twelve months."

"What other medications?"

As the talk turned medical, Paul decided to walk around the apartment. He started with the kitchen. He noticed a baseball bat behind the cabinet, next to the wall. It had a mitt on top of it. He stepped out and walked to the living room. On the wall he saw a reproduction of M.C. Escher's Crabs.

After another couple of minutes of exploration, he stepped back into the bedroom.

"Uh-huh," Durand responded, jotting down some things in his notebook. "Fine. I'll be back later. When's a good time?"

"Around twelve o'clock," Kendrey answered, collapsing onto his bed.

"In the meantime, all the meds are in the bathroom?" asked Durand.

"Yeah, and in my ass."

"Sure."

Durand went to the bathroom and pulled out the three bottles, taking four tablets from each.

"Bye now, Josh. I'll see you at twelve," yelled Durand.

"Shut the door on your way out," Kendrey answered.

"What's up?" Paul asked, as he hit the 'L' button in the elevator.

"The right drugs, the right doses. But he's despondent . . . something's wrong."

"I'm not an expert on these things, but could he have developed, what did the Mayo Clinic call it? . . . Yeah, a resistance."

"No. He shouldn't be in this condition, unless he's lying about what he takes."

"Yeah, but he's meticulous. Did you see the flat?"

"Mm-hmm. Might be fake drugs—you know, counterfeits imported from Asia."

"So they could look the same as the real product, but contain placebo?"

"Yeah. I'll run a few pills by the lab."

"So, when can we meet back here, eleven-thirty?"

Durand looked at him. "You're a real slave-driver, Robinson."

Paul chuckled.

Durand half-smiled. "Okay. How about you come back here at twelve-thirty. I'll come earlier to try and stabilize his condition. Otherwise it won't be worth interviewing him."

"Gotcha. Half-past-twelve it is."

Even with the three framed eight-by-eleven pictures of himself with the Monks at the Thiksa monastery in Ladakh, India, Paul had only half-a-box of personal items as he left his office. Mazurski ran into him.

"I was waiting to see you," Mazurski trumpeted.

"Does that mean you're buying lunch?"

"Not today, but make that a rain check."

"So, why were you waiting for me?"

"I've got two pairs of tix to the Penguins match this weekend. You and Latanya wanna join us?"

"Sounds tempting. But I think Tan has something going on . . . maybe it'll be just me."

"Then that'll leave one ticket. Give me a call by seven."

"Fine," Paul fist-bumped him and waved good-bye.

As he drove home across town, he considered his next steps. He took out his iPhone and pressed the 'record' button. After interviewing Kendrey, he would meet with the election commission within the Department of State. They had made the decision to allow these types of machines to be used in more than a dozen districts. If it was just the old folks' technical deficiencies after all, State would have had all the study results.

He speed-dialed the number given him by the Department of State for the woman who managed the random test of voting machines used in Pennsylvania. He was wondering how many machines were tested, as the phone rang.

"Maria Perrino, Department of State Voting."

"Hello, Miss Perrino, Paul Robinson, from WQHP here. I understand you organize the testing of the voting machines."

His query was met with a full five seconds of silence.

"Yes, that's true, but I believe it best that you speak with our media communications department spokesman."

"Sure, but maybe you can summarize—did you find any discrepancies?"

"Media communications spokesman Ian Smith will be pleased to answer your questions," Perrino said abruptly.

Paul heard only the tell-tale "click" of the phone hang-up in response to his question. *Not very nice.* He subsequently pondered where he would have to go to follow up on that angle. Then, abruptly, he changed thoughts. *Why did Leboeuf pull a gun on me?*

Before Paul could answer his own question, he had driven in front of his house.

Upon reaching the front door, he hoisted up his box-of-deskly-possessions, pressing it to the wall next to the doorframe, so he could open the door. He fished his key out of his pocket, and as he moved it towards the lock, he paused. The door was ajar.

He pushed the door open. *"Tan?"* he yelled into the space. No answer.

He quietly lowered his box and positioned himself squarely in front of the door. He took his Leatherman Wave multi-blade out of his jacket pocket, pulled the longest blade out and locked it. He kicked the door wide open, extending the knife in front of him.

The entire living room floor seemed to be covered three inches thick with his belongings. His suits from upstairs were strewn all over the chairs. Drawers from the den had been removed from the chest and emptied onto floor. He took his cell phone and speed dialed.

"*Hello?*" Latanya answered on the other end.

"Damn you, Tan', you didn't have to do this."

"*Do what?*"

"You know friggin' well what! I said take your stuff. You didn't have to trash the place. What, you couldn't find the—" He noticed Latanya's stuffed giraffe on the ground and paused.

"What, what are you talkin' about? I was gonna go tomorrow after—"

"Tan, wait . . . I think, I think we've been robbed. I'll call right back."

He walked to where his file cabinet normally stood. The cabinet was lying on its side. Files were scattered all across the floor. *Strange*, thought Paul, *the flat screen TV is still up . . . what in the bejeezus is this? No self-respecting robber would have passed up that state-of-the-art model.*

He opened the files one-by-one to see if any information was missing. They appeared intact.

Below a few files were his T-shirts, still folded. After a few more minutes of searching, he decided to leave everything as is; it would be easier for the police to get fingerprints that way. He noticed his second wallet, his 'mugger's money' wallet on the floor, next to the stapler. He recalled having left it in plain sight, on top of the printer. He squatted down, picked it up and flipped it open. It was missing the three twenties he had put into it the week before. He glanced up—his desktop computer was still there, untouched.

He stepped to the kitchen, rubbing his jaw, when his iPhone buzzed and dinged. He looked down to see an SMS had been received.

"*Pills were fake. Brot real ones. C u in 40.*"

He looked up at the wall clock. It was 11:50. "Shit."

He quickly shuffled through his shirts, his socks, his paper files and open photo albums. *Where's my laptop?* There was no time to clean up, either. He'd report this later. As he stepped outside, he tested the door lock. It was working. It had been a clean tumbler jerk, not a crowbar job. He couldn't help thinking how strange it was; a thief professional enough to open the door like a locksmith, and disarm the alarm, but stealing only sixty bucks and the laptop?

On the way to Kendrey's apartment, he dialed Latanya.

"Tan, we *were* robbed."

"Man, Paul, I'm so sorry. Was anything big taken."

"Of the good stuff, only my laptop. Tan, sorry for raising my voice."

"That's okay."

"Why the hell did the robbers need to throw a stuffed Giraffe too? Barbarians."

Latanya chuckled. "Paul . . . " Tan paused.

He understood, just from the tone of her voice. She didn't need to say anything more. But the words continued. "I hate to say this, but I'm still gonna take my stuff."

"Tan, I wish you wouldn't."

"Paul, it's just that things now . . . " Latanya was sounding strange. She was usually straightforward. This was apparently as hard for her as it was for him.

Suddenly, she changed her tone. "Listen, Paul, I'm at Penn, but I'll come by tomorrow. In the meantime . . . " Paul sensed she was smiling, "I'll keep my fingers crossed for you."

A couple of minutes later, Paul's phone rang. It was a number he couldn't recall.

"Hello?"

"Hi, Paul, it's Dave."

"Oh yeah, Dave. How you doin'?" Paul started, absentmindedly, still pondering Latanya's last sentences.

"Just wanted to thank you for teeing up the shooting range the other day, mate."

"No problem. Nice meeting you."

"Also wanted to say, that you might have been onto something. You know, as part of my work on the gas shale, I read through the press," he started, cheerily, "Anyway, this morning I read something in the Altoona Bend about a bloke who claimed the machine changed his vote. He was somewhere in Blair County."

"No kidding."

"Yeah. Anyway, it's on page two, at the bottom."

"Thanks for the info, Dave." Paul didn't want to appear rude to the person who might just have given him a first-rate lead, but he was going to have to maneuver between two cars apparently driven by geriatrics on sedatives. "We'll have to get together again soon."

"Yes, maybe at the range again."

"Yeah, why not? I'll call you Monday."

Paul felt like he'd just had a Vitamin C booster shot. He swerved the car to the right, around the car in the rightmost lane. A minute later, he pulled up to the curve and ran out to Nassim's newspaper stand. The guy carried every paper in the State, and when Paul had first moved here eight years back, he loved to read his own local paper. Now he burst into the store and scavenged the racks for the Altoona Bend. He found a copy on the bottom right rack and opened it to the second page. '*Are E-Voting Machines Reliable?*'

He scanned the article quickly. ". . .Stoltzfuss, a long-time inhabitant of Millersburgh . . . voting machine changed vote . . . after he had corrected it."

10

The curb in front of Kendrey's building had no spaces, so he parked around the corner from the building. As he exited his car, Paul again weighed calling the police. The correct thing to do, but then they'd pose tons of questions. Not generally an issue, but time-consuming. He figured that this evening would be best for that and put back his phone. In the meantime, he'd get whatever information he could from Kendrey. This afternoon, he would drive to Millersburgh to talk to Mr. Stoltzfuss about the alleged machine malfunction.

The more he reflected on the question of retirees voting, the more he considered that it might just be some older folks who hadn't grasped the technology. His own father drummed up dozens of excuses to avoid using the internet. The best one was "there's no place for the computer stand in the house," as if one needed a dedicated server cage the size of a tractor to house a desktop computer.

Paul looked at his watch. He still had five minutes. He'd better set up an appointment with Stoltzfuss. If he couldn't get a listing on the White/Yellow Pages app, then he'd have to start calling a whole string of contacts in the area. He downloaded a number from the app and dialed it.

"*Hello?*" a crusty voice replied.

Yes. "Hi, my name is Paul Robinson, I'm the reporter from WQHP in Harrisburg."

"Yes?"

"And I just read about your complaint and story in the Altoona Bend."

"Yeah, they changed my vote."

"Well, that's a subject of great interest to me. I was wondering if you wouldn't mind my asking you a few questions about it."

"With TV cameras and all? No, no, I wouldn't want to make a fuss . . . maybe I just made a mistake, like my wife says," the man replied.

"How about without TV cameras? You see, you're not the only person who's mentioned that the machine has changed your vote. But I'm interested in the details. It may be nothing, but it may be that the machines have problems."

"Well, all right. But no cameras, right? I'm a couple a' weeks overdue for my haircut," Stoltzfuss added.

"No cameras," Paul responded, already calculating how long it would take to get there. It would normally take two hours to get to his parents and another twenty minutes to go north. And that was after spending a half-hour with Kendrey. *Didn't seem this programmer could withstand more than that, given what all those books said about short attention spans.*

"Would four o'clock this afternoon be convenient?"

"Well, if it ain't too long. We got a prayer meetin' to go to at five o'clock."

"Hmm. How about I get there at a quarter to four."

"Sounds good. Oh, ya drink coffee or tea?"

"Either'll be fine, Mr. Stoltzfuss. Bye now." *Done.* Paul noticed the battery gauge on his iPhone at one-eighth strong. He bent down and glanced into the shelf above the ashtray. *Rats. Must've left the charger in the WQHP van Tuesday.*

He followed a tenant of the Executive Suites inside through the front door and to the elevator bank. The tenant got off the elevator on the third floor, while Paul rode it to the fourth. He got off and strolled to the door of Kendrey's apartment, then glanced at his watch. It was 12:32 – he was almost on time.

He opened the unlocked door and strolled in. He closed the door gently behind him. As he took three steps into the living room area, he noticed red spots on the floor. He stopped.

A high-pitched male scream.

Paul felt a chill go down his spine.

Kendrey.

"Leave me alone. Stick it up your ass!" Kendrey bawled out at the top of his lungs.

Paul pocketed his moleskin notepad and squatted down. He quietly slipped off his shoes.

On the table five screens, two with chess games. He took five steps and froze. Not far from the table was a pool of blood. Two tracks extended around the corner to the bathroom.

"So, you're not gonna take the pill?" Paul heard an unfamiliar voice asking Kendrey.

He took a step towards the kitchen. Then another, silently moving closer to the bedroom. Suddenly, he recalled the baseball bat. The kitchen would be obscured from the room, unless someone was standing in the closet.

Another cold chill ran down his spine when he heard Kendrey scream: "I'm not taking it!" followed by a discombobulated "I feel sleepy… what did you give me? Is that Citalo-Citalopram?"

I'll have to take that chance.

"That's a good Joshua," the stranger said, in a tone Paul found grating.

He tiptoed toward the kitchen. He reached the wall at the end of the counter, took the black mitt off the aluminum bat and laid it on the counter. He grabbed the bat by the sticky grip. It felt heavy, solid. Paul crouched down, panther-like. He tightened his hands on the bat handle. As he lifted the bat up, it clanged against the side of the cabinet. *Shit.* He swung the bat to his shoulders, before realizing how useless the gesture was. He expected the thug to walk out now and drill him.

But only Kendrey's gentle humming broke the silence. He relaxed his grip on the bat slightly. *Whew. They hadn't heard.*

Two cautious steps toward the bedroom.

"Fire, fire, bang, bang," the stranger sang, butchering some song.

Paul stopped next to the door. He could feel his heart pounding wildly. *Even if the guy was armed, he'd be facing Kendrey on the bed . . . holding meds.*

Paul's hands felt tense. No more time. He rushed into the room.

A man in a dark raincoat and gloves turned around, glanced at him and whipped out a gun with his left hand. Too late.

Paul brought the bat down on the thug's upper arm, extended to ward off the blow. The bat landed two inches above the elbow. "Cr-a-ck."

The gun fell on the bed.

The stranger recoiled, wincing. Paul lifted the bat, but the man had reeled backwards, turned his head to the corner and vomited.

Paul glanced towards the gun. It wasn't on the bed anymore. It was in Kendrey's hand. Kendrey held it loosely, pointing it at him, registering no emotion. Paul wondered for a split second if he should take the gun from Kendrey. *No, beat the guy first.*

The intruder grabbed the metal nightstand lamp with his right hand. Paul swung again. Loud bang. The bat hit metal. The lamp bent, but stopped the bat. Paul pushed with all his weight. The man stared at him with his olive, steely eyes. He kept the lamp perpendicular to the bat. The thug couldn't hold him much longer.

Suddenly, the man sidestepped, pushing the lamp, parrying the bat away from him.

Paul almost fell as the bat swung to the wall. As the man dropped the lamp he caught Paul's neck with the same hand. The lamp crashed. The man squeezed.

Paul felt the stitches of the intruder's gloves digging into his Adam's apple. His lungs were being closed off. Incredible pain. Blood rising to his temples. No air. *Kendrey's doing nothing. Doesn't he get it?* The room, the bed, started getting fuzzy.

He summoned up all his strength to send a kick between the man's legs. His foot bounced off the intruder's right inner thigh, but still found its way to the man's testicles. A pathetic attempt to emulate the karate forms he had learned as a teenager. But it worked. The man gasped, loosening his grip on Paul's throat.

Paul coughed and leaned against the wall. He tried once again to kick between the man's legs, but missed, instead his foot sailed far right of the man's crotch, and hit the man's broken forearm.

"Oww you fu—" the man yelled, bending down. "Shoot him!"

Paul glanced at Kendrey – he appeared to grasp the gun handle tighter. *Who would shoot me?* Only crazy Kendrey was here. *Is there an accomplice? No sounds....*

The thug fell back to the wall. Out of nowhere he flashed a switchblade.

71

Paul raised the bat high. The man lunged at him with the knife. Paul side-stepped to the left. *Not fast enough.* The blade sliced across his coat, nicking his chest. *No pain . . . yet.*

The man stared at the bat. Paul gritted his teeth, unleashing another side kick to the groin. This time, dead on.

The man doubled over, still holding his switchblade.

Paul brought the bat down on the thug's head. Solid. The man collapsed to the floor.

Paul stood above him, his hands clenching the bat handle sweating.

The thug didn't move.

Paul lowered the bat. He sat down on the bed and took six deep breaths.

"Bang, bang, fire, fire--" Kendrey gently sang out.

Paul shook his head. That guy on the floor was going to kill him. *What to do first?* Paul looked around the room, lost. He was still out of breath.

After another minute, he got up and raced to the bathroom adjacent to the living room, hoping that the puddles of blood he had seen on the floor were not Durand's. He slowly pushed the door open with his fist.

Durand lay still. He appeared to be staring at the ceiling. One bright red hole decorated his forehead, like a *bindi* on a fast asleep Hindu wife, while his chest area was covered with a dark red puddle three inches in diameter. Another puddle on the floor extended six inches beyond the body on both sides.

He knelt down beside Durand. *Oh my god . . . and it was my fault. Why him?*

Suddenly, he felt the blood leaving his cheeks. He leaned back on his haunches. "Evidence, evidence. Think *CSI*," he muttered to himself. He ran to the kitchen, picked up a glass saucer, pen and notepad that were lying next to telephone, and put them down near the sink. He picked up plastic ballpoint pen, took out the ink cartridge and with a knife from the knife block, cut the plastic ink cartridge in two. He shook one half to empty ink into saucer, but none came out.

"Shit!" Paul surveyed the kitchen in three seconds, noticing some yellow latex kitchen gloves next to the sink. He put them on and wiped down the pen with a sponge. He picked up the saucer and the knife and ran to the bathroom. With the knife, he scooped blood from the floor to the saucer. After four movements, he had enough of the liquid in the saucer for what he needed to do.

Keeping the plate steady, he ran back to Kendrey's room. He dropped next to the intruder. He ripped off the man's right glove, and dipped his thumb in the saucer. He pressed the fleshy tip against the notepad paper. Paul took the man's index finger and repeated the exercise.

"Happiness - is - a fast gun," Kendrey sang out weakly. Paul looked up at him. Kendrey was aiming the gun at his chest. "Shoot, shoot . . . kapow," he sang.

"Kendrey, put it down. You've put your fingerprints on it. Just put it down, before you take all of this guy's prints off it."

Paul put down the notepad and scratched his head. Suddenly it felt hot. He was sweating profusely on top as well. He stepped to the nightstand and took a tissue from the box. He wiped all blood from the yellow kitchen gloves.

He took the man's limp index finger and once again dipped it in the blood and shook the excess off. Paul lifted the body up and dragged it next to the bed. He lifted the limp thug's right shoulder and imprinted the teak-finish headboard with the man's index finger and thumb prints.

"You're making a mess of my r-room, my headb-board!" Kendrey blurted out.

"Sshh Joshua, this is in case they try some clean up. Chances are they'd miss the teak – the CSI team won't."

Paul dropped the body, and examined the prints from two feet away. The blood was barely noticeable. To the sweeping naked eye, invisible.

Paul slid the man's fingers back into his glove.

He turned the intruder face up and patted down his suit, stopping above the right pocket. He reached in and slid out the mobile phone. He pushed the calls log button until the 'Received Calls' signal lit up. As the number came up, he transcribed it – 9 '983-456-884'

On the notepad, next to the number, he wrote as legibly as possible -- 'FINGERPRINTS. 'INDEX FINGER', 'RING FINGER'."

Paul pushed the buttons on the intruder's cell phone, and the operator came on in four rings. "*Yes?*" she answered lazily.

"Need to report a shooting, a murder."

"*Who's calling?*"

"The address is forty-one Second Street, apartment four-twenty-two."

"*Identify yourself please.*"

"I'm the killer."

Paul hung up.

The thug stirred, shaking his head.

No way, asshole. Paul grabbed the bat and whacked the man over the head just as he lifted his head. It was a hard hit. The man's head dropped limp again.

Kendrey sat up in bed, looking at the killer.

"Joshua, we've got to go for a ride now, a road trip," Paul stated, smoothly.

I don't want to go anywhere."

"Josh, we have to go, or else they'll kill you."

"I don't care. Let them."

"Josh, these guys were swapping your meds. You were taking blanks."

"Really?"

"Really."

"That's not very nice."

"Come on, let's go before the police get here, otherwise they'll arrest you. And me too – my fingerprints are all over the place."

"No, *no*, I hate the cops."

"Fine. Let's go someplace safe. Let's take the road-trip. I've got to talk to someone who lives in a really fun spot."

"Well, if we're going on a *road trip* . . . " Kendrey parroted back. He jumped off the bed. Suddenly, manically, he ran out of the room, to the kitchen. Paul heard him opening cabinet after cabinet.

"We'll start with Granola bars. All that yummy goodness. And *healthy* too!" Kendrey yelled out.

74

Paul followed Kendrey out and saw him rip open one granola bar package. The guy stuffed half a granola bar into his mouth. He chewed wildly as he started cleaning out the cabinets.

Paul realized he had one more thing to do. He turned around, hearing items fall from the cabinets onto the counter. His gloves still on, he lifted the mattress, and put the gun underneath it. He also ripped off one of the two sheets of notepad paper with the man's fingerprints next to the gun, under the mattress. He left one sheet on top of the bed. He took the third page of the notepad, folded it and put it in the inside pocket of his jacket.

"Joshua, could you please hurry. The cops'll be here any minute."

"Yessir. Just a couple of e-dibles and I'll be all set."

Paul shook his head. "Okay, but fast. I'll pick up the meds Durand brought." He picked up fifteen of the pills scattered all over the living room floor. He went to the bathroom to get the other bottles.

"Total breakfast for a great feeling in the morning!" he heard.

Kendrey had his head fully in the refrigerator when Paul returned to the kitchen.

"Boar's Head. Mmm, salted dead pig is my favo-o-orite," he sang out, before stuffing a slice in his mouth.

Paul entered the kitchen, holding the bottle.

"Joshua, are you ready?"

"Oh yes. But you wouldn't dare go anywhere without Mallomars, would you?!"

"Joshua—We've *gotta go*. Come on!"

"Soo-oo nervous. You have got to take some of my tranquilizers."

Kendrey tossed all the food products into two brown bags he pulled from under the sink. "I had these green and yellow ones, and *wow*, life was one Tiramisu cloud frothed up with a steam nozzle." Kendrey picked up an open bag of potato chips from the bottom and, as he lifted it up, potato chips scattered in all directions.

Paul couldn't stop himself; he grabbed Kendrey by the shoulders and shook him. "Normal!"

"Be civ-civilized! I'm a programmer of Danabold Technologies. Hands *off* me."

Paul looked in Kendrey's eyes. He detected a lost look, and unhanded him.

"Lead programmer, I'll have you know, of the Election Erection Project. Up yours!" Kendrey said, thrusting his forearm upwards obscenely.

"Okay, okay. Just grab the other bag and follow me."

"Wait, wait. I've gotta call my Mom, in case she calls here."

"Okay, but hurry."

"Where're we going?"

Paul wondered what he should say. The police would be here soon and would try to follow him. Would he be able to get to Millersburgh in time to get the story, before the death here embroiled him with the cops?

"Just say East towards Philly."

"That's weird. She'll have a panic attack if that's all I say."

"Okay, okay. Berks County."

Kendrey speed-dialed from his phone. It rang a long time.

"Hi, Mom, gone to Berks County for the day, with the reporter, the one who does the human interest stories on chimps and Olympic swimmers . . . something about a guy who complained about the machines. I'll be s- s- safe. Love ya, too."

"Fine, let's go," Paul barked.

Paul grabbed one bag, the bat, and walked to the door. He opened it a crack. The corridor was empty towards the stairs. Silence. He stuck his head out and looked in the other direction, towards the elevators. One elevator was going up to the fourth floor, the other was on the second floor. "Let's take the stairs."

"I haven't had any aerobic exercise . . . sounds just peachy!"

Kendrey went first, and Paul shut the door.

He passed Kendrey and walked quickly down the stairs, reaching the landing on the second floor. A middle-aged male came up the stairs. Paul hid his face with the grocery bag. As he passed the resident, he bumped his shoulder.

"Hey!"

Paul heard the guy stopping.

"Hi, Josh. Where you going with that jerk?"

"Hi Ed, I'm going with my f- f- friend on this kick-ass picnic. Mallomars, softball and the whole nine yards."

Paul was already on the landing of the floor below.

"In November?" Paul heard the neighbor asking. "Whatever. Just don't kill yourselves getting there."

"We won't kill ourselves, we've done enough ki—"

"-cking ass in that Scrabble contest!" Paul shouted. "Josh, I still don't get how you came up with that eighty-pointer, 'incarceration'!"

Paul heard Kendrey following him down another flight of stairs.

"Boy, you're dumb, everyone knows you can't get an eighty-pointer with that, unless someone had 'ration' before—but there's a one to fifty-three thousand chance—"

"Josh, we're late to the outing already."

Once outside the building entrance, Paul held his hand up. Kendrey stopped.

Paul looked both ways in the street. He saw a flashing police light some two hundred yards down the road. "Follow me."

Once at the car, he opened the back door, dropping the bag and the bat in the back seat. Kendrey went around the other side, opened the door and threw his two bags into the same seat. Popcorn and chips spilled out all over.

"Jeez, Kendrey, can't you put things down normally?"

"Yes, boss. I'm just so excited!"

As he pulled the Fusion away, Paul noticed a black Explorer pulling into a spot on the other side of the street.

He now looked straight ahead as another police car, lights flashing, passed him. The fastest route would be the turnpike to 220. Then North until 164, when he'd cut over to Roaring Springs. From there, maybe another twenty minutes to the Stoltzfuss farm, depending if there was snow. *But there should only be a dusting in the Appalachian foothills at this point of the year, if any at all.*

"Now, Josh—we're going to the countryside."

"Sure. Boy, I love it, with Driver ants eating up the food, and you with it."

"Good. In the meantime, take your medicines. This is the real stuff."

77

Paul pulled two prescription bottles out of his coat and handed them to Kendrey.

He watched him open the first of the bottles, then turned his attention to the road.

The car passed the Soldiers and Sailor's Memorial Bridge. The two massive eagles perched atop the spans always gave him the impression they might scoop him up like some grass-nibbling rabbit, cracking his ribs with their red granite claws in the process, for the mere satisfaction of feeling him writhe.

11

As Antoine entered the apartment, he glanced to the right, then to the left. He took one step forward. He noticed the edge of the blood puddle on the floor and pulled out his Beretta. As he took one step towards the body, he heard grunting sounds and something that made the floor squeak.

He moved towards the bedroom, gun drawn, arms down. Gun first, he lifted and swung his arms into the space.

Depalma was wiping the floor with a wet kitchen rag. Next to it was a jug of chlorine. He looked up as Antoine pointed the gun at his head.

"That's enough, *Bourne*! Help me clean up," Depalma blurted out.

Antoine holstered the weapon. "Whose blood?" he asked.

"A mental health nurse. He discovered the drugs were fake."

Sirens wailed outside the building.

Depalma looked at Antoine. "Shit! Who called them?" he asked.

"The Girl Scouts, Prince Charles. It don't matter. We can't take the body," Antoine answered, "let's split."

Depalma tried to put the arm through the sleeve, but he snagged it. "Shit!" he yelled.

"What the hell happened?"

"Fucker surprised me with a bat."

"Okay. Just throw your coat over your shoulders. I'll take the stiff's wallet, it'll give us some time," Antoine turned around and bounded out of the bedroom.

Once in the bathroom, he lifted the corpse's hip by putting his finger through a belt loop and lifting it up. He deftly picked the wallet out of the dead man's back pocket. He felt the front pocket, and removed the cell phone and hospital ID cards.

He stepped briskly to the window and looked out. The police car had stopped and two policemen were crossing the street.

The front door would be closed, so the cops would have to wait until someone let them in.

Antoine stepped in front of his colleague, opened the door and peered out of the apartment, sweeping the hallway. He saw the elevator descending from the sixth floor, it had just hit the fifth.

"Let's go. Elevator!" Antoine ordered.

Antoine ran to the elevator bank, and lunged for the button. The elevator opened. Depalma slid in behind him. Antoine pushed the 2nd floor button. The overcoat was falling off. Antoine lifted it back onto Depalma's shoulders.

The elevator stopped. Antoine pushed the "door open" button. He waited. Soon he heard the policemen coming up the stairs; quick and determined steps. Antoine pushed the 1st floor button.

Antoine and Depalma exited the building. They ambled nonchalantly towards Depalma's car.

"I'll get mine later. I'll drive ya to the center in yours," Antoine said.

***

With each mile, Paul felt a deeper sinking feeling for having brought Durand into this investigation. But how could he have known that the stakes had been so high? For whom? *Who could be behind this insanity? Someone in the Pennsylvania Department of State? Danabold Technologies? The Democratic Party? The Republican Party? Maybe Kendrey was tied to drugs.* Paul glanced over at Kendrey *Nah.*

As the car passed Mechanicsburg, he contemplated other scenarios.

He couldn't believe he was leaving the scene of a felony crime. He then imagined the ideal scenario unfolding—the police finding the intruder and then arresting him on the spot. The thug wouldn't have had time to discard the weapon, even if he did find it between the mattresses.

As he approached the five mile out-of-town mark, Paul started wondering where they might stay. He had enough cash to pay for a Motel 8 but not more if he intended to eat and pay tolls. He had seen enough TV shows and movies to understand that if either the police or the criminals after Kendrey wanted to find them, they could do it through tracing the credit card transactions.

Kendrey huddled in front. "Can't place . . . " Kendrey's jaw locked briefly,

"Incarceration," he mumbled for the fifth time.

"You are so right. It was absolutely . . . infantile of me. Now, on a more important note, do you think we could talk a little about the voting machines?"

"I can't. Vow of silence. If word gets out, I lose my four—"

"I don't mean to burst your bubble, Joshua, but they switched the meds because someone wants you brain-dead!"

"Really? Who could that be?"

"I dunno," Paul added with exaggerated naivete. "I thought you might be able to tell me."

Kendrey appeared to contemplate the question. "Hey, what apps do you have on your iPhone?"

"Josh. Someone's trying to turn you into a zombie, and they've scared Martin to the point that he's pulling guns on reporters, and you're asking what apps I've got on my iPhone?" Paul burst out, failing to contain his irritation. He turned back to the gray road. "Amazing."

"'Amazing' would get you seventy-one to one-hundred-thirty-seven points, twenty-one to eighty-seven for the word plus fifty points for using all seven letters."

Paul shook his head.

"What about just game apps?"

"For you, Pee Monkey and Moron Test."

"I've played both. Pee Monkey is stupid. Moron Test is cool, but it only took me twenty minutes to get through it. Just logic. What else do you have?"

Paul grimaced. He'd had an easier time with chimps. "I'm not going to give you my iPhone for anything, unless you talk."

Kendrey looked out the window. He looked at the trees, he looked up.

"We're not going to Berks County, are we?"

"No."

"I just lied to Mom?" Kendrey blurted out, before putting his hands over his ears. He started shaking backwards and forwards. After the third repetition, his movements became violent. Kendrey hit the dashboard with this forehead.

"Okay! I've got Doodle Jump, Pocket God, Angry Birds, Bejeweled!"

Kendrey stopped shaking.

"Pocket God is good . . . I've got to call Mom first."

"Fine, fine," Paul countered, just wondering how the great Almighty could have created such a dysfunctional loser and genius in one persona. He pulled out his phone and gave it to Kendrey.

"So, where are we going?"

"East Appalachias."

"What city?"

"City? Try 'town'. Whatever . . . Altoona," Paul answered, upset that he was being played by this freak of nature. This guy's mother, probably in Ohio, wouldn't have a clue where Millersburgh was. Elbow-grease Indiana. Wisecrack, Montana. Millersburgh, Pennsylvania, would be the same. At least Altoona she could find on a decent regional map.

"Why?"

"Because of an article," Paul summarized.

"Wha-what article?"

"An article . . . doesn't matter. A guy that says his vote was changed."

Paul heard the phone click on.

"Hi Mom! . . . I'm fine. No, I'm not at home. I'm with the reporter . . . yeah, this really cool one who does human interest stories on WQHP . . . We're on our way to Altoona . . . I'm not sure why. Yeah, he-he's a good driver, very g-good," Kendrey stammered, looking at Paul. "I feel s-safe."

Gray clouds above threatened rain . . . or, at these elevations, sleet. The coating of snow on the road thickened as the snow began falling harder.

How am I going to approach Latanya? She was proud, fiercely so, but why did she have that outbreak? Then it dawned on him. Wasn't she approaching thirty? Was she thinking "children?" If they were going to have kids, he would have to provide for them and, at this point, he reflected, he was really an overpaid adolescent.

12

Kendrey tapped the iPhone mercilessly. Apparently, he liked to see the pygmies dance.

The normal meds seemed to be kicking in.

"Glitch?" Kendrey suddenly asked, without provocation. "What glitch?"

"I thought you'd tell me. Something about the machine changing votes."

Kendrey shrugged his shoulders. "Leboeuf is okay, but he can't program betas worth a used Superman suit. Actually, I think Duprix programmed that qu-quadrant . . . Duprix knew his stuff."

"Duprix? Who's Duprix?"

"Another guy who worked on the program. Has a b-brain at least. Has a fast car, too. Wants to take me for a ride at 100 miles an hour. Anyway, Duprix d-did the switch, and they didn't l-let me reconfigure it f-for the final check. Since it was a five-hundred-twelve-M-B key, any Tom, Dick or Harry could have done it. Okay—anyway, they wouldn't let us talk to each other or even see what the others w-w-were doing. The boss said they'd miss delivery. Th-that's all. Can't tell more."

Paul started getting angry, when he saw a number of red brake lights up ahead and a traffic jam. It didn't take him two seconds to realize that it was an accident of some kind. He turned on the radio. "Wait, Josh, you can't stop there, but wait for now . . . I've gotta find out what's going on here."

" . . . *More on the turnpike pile-up. Reporters getting there . . . appears a tractor-trailer jack-knifed,*" the weather and traffic channel blared.

Rats. It's a gamble now. Paul started calculating what the chances were.

" . . . *Approximately twenty cars collided . . .* "

Paul recalled when the pile-ups took place on the interstate, it would take upwards of two hours to clear them. Right now, traffic was light, but this would change quickly. It would be faster to take the old Route 30, if he could get there without making a long detour.

The sky, clear until now, appeared dark and menacing, as the first snowflakes appeared; they fell in a swirling pattern, thrown in many directions by a rapidly changing wind.

Paul glanced at the car clock. The light blue LED read 2:41. One hour since leaving town. He looked up and saw a sign: 'Sideling Hill Rest Stop.'

The cars were now slowing to ten miles-an-hour. He pulled onto the shoulder and pushed the gas. Less than a half-minute later he was looping into the enormous rest stop. As he turned in, he looked up and saw three helicopters hovering over the area. *Must be massive.*

"Big pile-up, so we're gonna take a different route, Josh." Paul looked at the fuel gauge. "Besides, we're low on gas."

"Low on gas? What, do w-we have to go somewhere else after checking out the article g-g-guy?"

"Maybe."

At the Sideling Hill Rest Stop, like most rest stops on the new Pennsylvania Turnpike, the core structure was a streamlined operation designed to accommodate thirty truckers and at least three hundred hungry, thirsty, bladder-bloated or confused drivers and clueless passengers at the same time. By way of a purpose-built overpass the stop could accommodate cars coming from either direction, while keeping the westbound vehicles separated from the eastbound ones.

Paul drove up to the gas station. Many others had the same idea. He steered the Fusion behind two cars waiting to fill up.

"Josh, if you wanna take a leak too, now's the best time to do that."

Kendrey appeared confused. "Where else are we going after the article guy?"

"To someone else that may have had D-R-E problems. Just go on," he said, reaching over Kendrey and opening the door.

"While you're at it, get me a coffee too. Double latte, cinnamon flavor," Paul said, handing a 20-dollar bill to Kendrey, "and get yourself something too, we're gonna have a long haul ahead of us."

"Want me to charge the pho- phone a little, too?"

"Good idea," he said, handing his iPhone and the socket charger to Kendrey.

Kendrey set off quickly to the building.

Paul studied the 'barrier' between the building and the cyclone fence two hundred feet to the northern side. At one end was a gate. Locked. Between the gate and the main building was a sidewalk, slightly higher than standard. The Fusion wouldn't clear it easily, but it would clear it. But on top of the sidewalk, about twelve feet apart were yellow metal posts, connected by a yellow plastic chain. Even if Kendrey were to hold up the chain, the Fusion wouldn't fit under it. A 'PA Turnpike' maintenance truck stood next to the barrier on the other side. *No good.*

He advanced the Fusion as the next car departed. He looked to the westernmost edge of the exit ramp. There were no barriers after the ramp ended and before the fence started, a space of about twelve feet.

A pump freed up and he pulled up next to it. As he squeezed the 'premium' gas button, he heard one of the helicopters. Two were above the traffic, a little bit in the distance, while one was circling directly above the rest stop. It was from Philly . . . WZTX. Big distance to travel for this. *Maybe doing a story on the turnpike.*

His mind drifted back to Durand, staring, lifeless, at the ceiling. He would have to phone Elaine at some point, to tell her what happened. She had to know why he had left without explaining to the police.

Paul suddenly felt the blood leaving his skull, as a wet chill descended his spine. *Oh sheisse! How could I have forgotten his phone?* Whoever found it first, would know that Durand's last call was to him. If the cops didn't find the evidence before they found the phone, he'd be in trouble. In fact, the police could be in pursuit of him now.

Paul collected the receipt, drove the car forward and waited, planning his exit. *Back roads . . . thank God I know them.*

Five minutes later, Kendrey returned.

"Thanks much, Josh. How was it inside?" Paul asked as he removed the pens and notepad from the rear cup holder and put them in the glove compartment.

"Crowded. I hate crowds of people. They smell."

"Can't argue with that. Now, Josh, I'm going to ask you to put your coffee in the holder for just a couple of minutes and leave the lid on. We're in for a little dare-devilish but time-saving maneuver. Also, take the chip out of my phone."

The parking lot was now full as he guided the Fusion to the gap he had scoped out. He saw, in the distance, two more police cars and two more fire trucks make their way west through Sideling Hill, which had been sheared some forty years earlier to accommodate the new eight lane turnpike.

"Kendrey, you strapped in?"

"Yeah."

He drove up and over the curb.

"My coffee!" Kendrey yelled, as the Fusion tipped to a 20-degree angle. "Off road?! You can't do that!"

Saying nothing, Paul drove down a bank, until he came to a gravelly road with a dusting of snow. *And I don't have my winter tires on.* He ignored the programmer's rant, as he calculated there would be about another hour and fifteen minutes of light. But it felt, at this instant, that darkness was imminent.

Paul drove along the now-white gravel road, past a number of farmhouses. After a half mile, this "Postal road" took a slight grade. A few minutes later he would be on Route 30. Suddenly the Philly helicopter appeared above the treetops and hovered in place for a couple of seconds. *Was there an accident on this road, too?*

13

"So Josh, you were saying that the boss was afraid you'd miss delivery." Paul gently guided the car over the gravel, around potholes, real or perceived.

"D-d-don't want to talk now. I hate c-country roads. I would've told Mom."

"Come on, Josh."

Consumed by the silence, Paul started contemplating how he would find Duprix, the third programmer, tomorrow. He could talk fast cars, one of his hobbies, ad infinitum, if that was what floated Duprix's boat. Twice per month he had an 8:30 slot, virtually the crack of dawn, reserved at the Williams Grove racetrack. A half-hour of driving at 120 to 140 miles an hour got his adrenaline pumping mightily. A few times, to break the routine, he even drove backwards at eighty miles-an-hour. He could tackle the toughest stories after that driving. It also connected him to his inner boy, to the joys he felt going down the all-wood roller coaster at the Mechanicsburg amusement park.

Could this Peter Pan complex explain why I have this underlying fear of commitment? He swerved slightly to avoid a squirrel that had stopped on the road to enjoy an acorn meal.

Paul heard a popping sound above him. Some object perforated the lid of the coffee cup in the first of the two holders. He looked at the cup. *Meteorite?*

He heard two more punching sounds at his door. He glanced at the roof, felt the brisk draft. He jerked the steering wheel to the right. *They were being shot at.*

He jerked the car back to the left.

"Stop it, you'll spill my coffee!" Kendrey blurted out.

Paul swerved from left to right, ignoring Kendrey, as slugs peppered the snowy shoulder of the oncoming lane, throwing up wisps of snow. He saw the treetops swaying as well. *The gusts must make it hard for the copter to move--.*

"D- dude! Don't drive crazy, you'll make me sick," Kendrey bellowed.

Paul didn't hear Kendrey's further comments, as he looked out through the passenger window, up and behind him. He saw the helicopter bobbing. It tried to swoop down. He slowed the car.

The helicopter shot forward and up slightly, then hovered. Paul floored the accelerator. The Ford swished past the slowing helicopter. As he did, he heard a 'pop' from somewhere. He looked in the rear view mirror. A plume of smoke five feet high rose from the left side of his lane.

The bottom of the helicopter landing skids accelerated forward.

Once again, the helicopter lurched to his car's side. When the helicopter cabin was even with the Fusion, Paul hit the brakes again.

But this time, he saw an oblong object fall out of the helicopter window about fifty feet in front. The lane flashed a bright orange. He jerked the steering wheel hard to the right. The Fusion swerved, fishtailing on the road, then rode down the shallow ditch and up the hill. Paul cringed as the fender hit a rock.

The car shuddered.

Fender's broken. But Paul managed to get the car back onto the road, still traveling at 50 miles an hour.

A car emerged from around the bend in the other lane. The helicopter lifted up.

They're not risking collateral damage . . . or witnesses . . . Paul glanced in the rear-view mirror. There were no cars behind him.

An idea took hold. A risky plan, but if he could evade the helicopter for another three-four minutes, it could work.

The helicopter still hovered ahead of the Fusion, as the oncoming car passed him.

He saw a long object with a Christmas-light-shaped bulb on the end of it emerging from the right side of the helicopter, which was turning slightly. *RPG.*

"Brace yourself!" He pressed the accelerator and started a swerve pattern.

Three seconds later, a bright flash appeared next to his door. The ground shook. The car lifted to the right. What sounded like a hundred stones and a big rock hit the left side. His window cracked. His left calf burned. Paul struggled to keep his foot off the brake as the car returned to the road, gaining speed.

The helicopter was now so far back and to the right that it was out of sight. He glanced at his rear-view mirror. Nothing. But he passed a road and a car was signaling to enter, though it was headed in the opposite direction. Another car appeared over the hill.

Who? Why?!

14

Halfway there. But have to evade the helicopter for just one more minute.
Paul had driven Postal Road 248 dozens of times in his teenage
years, when he spent summers with Uncle Pete, but never under
duress.

It was just over a mile from the Turnpike to the unmarked
road that led, a hundred yards later, to the barriers blocking off
the old turnpike road. Paul caught the small feeder road on the
right.

The barriers – standard 8-foot long, 32-inch high cement
highway siding pieces – were placed three deep, and not in rows,
but interspersed, like bricks in a wall. Bikers would wend their way
through these, to get onto the old interstate.

Paul jumped out and looked to his left. A fallen tree blocked
the space between the barrier and the five foot high rise. He
looked to his right. Not a chance, the concrete barriers went deep
into the forest.

In the distance, he heard the sounds of helicopter blades
chopping the air. *Shit.* He looked back to the left. If he could
remove the fallen tree, there would be space enough for the car.

"Josh, quick, I need your help. Come with me."

He dashed to the tree trunk, which was about five inches in
diameter, and lifted it up.

It lifted easily despite still boasting all its leaves, now brown
and damp on top.

"Josh, hold this tree up, I'll drive through."

Kendrey nodded, and lifted the tree up.

As he approached the car, he heard, more loudly, the 'tat-tat-
tat' of the helicopter blades.

Paul quickly drove the car under the fallen tree, until the trunk
was over the rear door. Now, the 'tat'-'tat'-'tat' sound was replaced
by thwack-thwack-thwack. He stopped the car, jumped out and
gathered some large branches from the ditch.

"Josh, now come help me cover the car," Paul yelled, over the now deafening sound of the approaching helicopter. They picked some more branches from the tree and placed them over the exposed parts of car.

Kendrey stepped back, it appeared, to admire the work.

Paul leaped forward, grabbed him by the arm and pulled him under the rotting tree foliage. The ground shook. The helicopter flew at fifty miles an hour, four hundred feet over the thinning orange and brown treetops.

Through those trees, Paul thought he saw the pilot pointing at what would be the Sideling Hill Tunnel.

The helicopter approached the turnpike. The two other choppers were within view. The helicopter then started looping to the East, away from them, away from where the tunnel would be.

Paul breathed a long sigh of relief. "Josh, let's go!"

He checked to make sure the lights were off as he drove the Fusion over the snow-brushed brown bushes and dead plants, which now constituted the central dividing line between the westbound and eastbound lanes of the old turnpike road. It had been closed to automobile traffic for nearly forty years.

Paul accelerated slowly as he entered the westbound lane. Not that it would have made a difference, no bicyclists or ATV riders would venture on this road in early November, an hour before dusk, during a low-grade storm. But at no cost did he want to signal the brake lights. *Just another three hundred yards.*

His left calf started aching. He reached down to rub it. Something was sticking out and he pulled it out completely.

He brought it up to see. A sharp metal splinter. He had never been chased down like, like a rabbit before. Without knowing why, he hit the steering wheel hard with his left palm. "Shit! Shit! SHIT!" he yelled loudly. He glared at Kendrey. "WHAT is going on?!"

Kendrey turned to him with a panicked look.

Suddenly, through the cracks in the window, Paul heard the tat-tat of the helicopter again. The helicopter was gaining on them! He slammed the brakes and pulled on the parking brake.

A projectile shot five yards past his left front, throwing up a fountain of snow. The snow and something metallic hit the

92

windshield, breaking it. The car turned ten degrees northward. Paul felt disoriented for a few seconds.

Finally realizing what had just happened, he released the parking brake and jammed on the accelerator.

"What the hell?!" yelled Kendrey.

Paul heard the long 'crack' as something impacted the bottom left section of his rear windshield. At the same time, another hundred rocks impacted his rear window. Glancing into his rear-view mirror, he saw the upper part of the rear windshield had dozens of cracks in it. He once again jammed the gas pedal to the floor.

Paul jerked the steering wheel two inches to the left, then to the right, as the Fusion raced into the east entrance of tunnel. Another grenade exploded some twenty feet in back of the car. Looking into the rear-view mirror, he saw the helicopter skids hovering on the outside of the tunnel.

He drove the Fusion about 100 feet in, holding the steering wheel hard to keep the car from drifting to the left. The car braked to a stop. Paul rolled down his window and stuck his head out.

The helicopter hovered low to the ground and turned parallel to the tunnel entrance. For a second, he thought the helicopter would land.

Instead, he saw the gunner waving . . . then smiling at him. Paul thought he saw the gunner mouthing the words "I'll be back," as he made a pistol gesture with his right hand.

The helicopter lifted off swiftly.

As he listened to the helicopter flying into the distance, Paul leaned back in his seat. He took a deep breath and exhaled. He turned to his right.

Kendrey was shaking back and forth, murmuring, "Glitch, glitch. But they're still assholes!"

"You wouldn't have any scotch in those bags, would you, Josh?"

Kendrey stopped shaking. "I don't drink alc- alcohol."

"Your loss."

Paul turned straight ahead, into the darkness. He turned on the lights. They flooded the next fifty yards of the tunnel. He turned on the high-beams and the light extended another fifty yards.

"They must want you real bad, Kendrey."

"Why me?" Kendrey asked.

Paul shrugged his shoulders. "I was hoping you'd tell me."

Paul looked Kendrey in his left eye, which now faintly reflected the light from the headlamps.

Kendrey stared back, emotionless.

"Anyway, we're gonna wait here for a while, until it clears."

"We're going to wait here?"

"Yep. I want them to either come after us, or give up waiting. If they come after us, you see how low and narrow this tunnel is? The helicopter won't be able to turn around or it'll have to fly so low it'd hit us," Paul said driving his fist into his open palm. "Besides, I gotta check the tires. One of 'em feels flat."

They both got out of the car. As Paul stretched out, he felt the sharp pain again in his calf. He lifted his trouser leg. The wound was less than a quarter inch in diameter, and small stream of blood it had generated had already dried up.

He walked to the back and checked out the tire. It was fine. He strolled back towards the front tire. Completely flat. He looked closer and saw a two-inch gash an inch from the rim. He returned to the back, opened the trunk and removed the spare.

"Hey, for a r-reporter, this was pretty smart. How'd you know about this tunnel?" Kendrey asked.

"Thanks Josh, I see tact is your strong suit," Paul said, shaking his head, "I grew up thirty miles from here, in Loysburg. My uncle Pete lived on a farm a few miles down the road that way," he said, pointing towards the tunnel entrance, so sometimes I'd come play with my cousins in here." He went to the trunk and removed the jack. "The funnest thing to do was to go in the middle and play Marco Polo. Couldn't see anything."

"Yeah, it's dark."

"Which wasn't bad for making out a few years on," he added, as he turned the handle a few times, lifting up the left side of the car.

"W-With girls?"

94

He cranked the handle slowly as the tire lifted up off the ground.

"No, with cows! Of course with girls. You know what the best line was?"

"No, what?"

"'See how dark this is? The eyes can only take forty-eight hours before the rods and cones inside become so exhausted from their work, that a person goes blind'" he said, as he took the spare from the back, "'Like I'm going blind with my love for you.'"

"That's just dumb."

"I won't say it isn't. But it worked oh, about a third of the time." He unscrewed a bolt from the flat tire.

When the tire was on and the flat put in the back, he invited Kendrey back into the car. The guy was shivering.

He started the car up and turned on the heater.

"I reckon they should be long gone by now, what do you think?"

Kendrey nodded.

As the car advanced, Paul glanced at the graffiti he had written 15 years earlier, two hundred feet from the western end of the tunnel. "Paul loves Linda and Linda's tunnel." He stopped the car. "See Josh, I even named it after one of my girlfriends," he said pointing to the graffiti.

"Perv-pervert!" Kendrey yelled. "Okay, I'll t-tell you what I kn-know."

15

The Fusion exited the tunnel and entered the relatively blinding gray daylight at 40 miles an hour.

Kendrey blinked several times. He squinted. "So, Duprix di-di-didn't catch that mistake," he said.

"Wait, Josh. Let me make sure this is still recording." Paul reached for his iPhone on the dashboard. He had decided to take the risk of being traced and so was going to put the battery back in. The info Kendrey was spewing out was just too voluminous . . . and technical. As he picked it up, the iPhone slipped out of his hand, bounced off the edge of his seat and off his left shoe before falling behind the accelerator pedal.

"Shit." Paul guided the car onto the gently bending curve and then reached down to pick up his phone. He groped for it. Nothing. The phone was wedged between the accelerator and the mat. With his index finger, he pushed it out a little. He needed the phone out another inch before he'd be able to grasp it between his thumb and index finger. To keep the car going, he pushed on the accelerator with his left hand — nobody would be on the road and that stretch was straight.

Suddenly three rapid punching sounds. Glass shattering. Panel exploding. Steering wheel vibrating against his ribs. *Hell, not again!!* Still hunched with his head just below and to the right of the steering wheel, he jammed the brake hard with his hand.

It seemed an eternity before the car came to a stop.

***

From the hill on that side of the tunnel's West exit, Jenkins dropped his scope down a fraction of an inch and saw the brake lights go on, while the car skidded to a stop, angling too far to the right to be a normal stop. Between the cracks in the back window, he could make out that Robinson was down in front, his body slumped to the right of the steering wheel.

"Got him," Jenkins shouted, as he put the gun down, and stepped gingerly back towards the helicopter. He hopped over some exposed roots, eager to see the corpse of the reporter that had almost derailed "The Program." He heard the pilot's door open.

"They're moving back into the tunnel!" Bueller shouted, gesturing towards the road below.

Jenkins swiveled around to see the Fusion had moved. It was going backwards, swerving from side-to-side and was now about fifty yards from the tunnel entrance. Jenkins once again took aim, but a copse of trees obstructed his line-of-sight. *Dammit.* He would have one shot, a spot just in front of the entrance. He brought the automatic rifle to his shoulder. As the Fusion fishtailed back into the tunnel, he squeezed off a burst of three rounds. One ricocheted off the concrete lip of the tunnel, perforating the front hood of the car.

***

As he maneuvered the rear of the Fusion into the tunnel, Paul heard the cracking of glass. His left hand exploded in pain. He gritted his teeth, determined to keep both hands on the steering wheel. He braked as the light thinned out, about seventy yards inside.

He turned his head back towards the tunnel exit. The windshield had a crack, extending from behind the right side of the rear-view mirror to the left of the steering wheel. In the middle of the windshield was a small hole. The angle of front windshield was almost 45 degrees, so this hole was a full foot in front of the steering wheel, but now his vision was compromised.

Paul pulled a glass shard out of the back of his left hand. He looked past his hand and saw a hole where the 50 once stood on the speedometer. He turned to Kendrey.

Kendrey was bobbing back and forth, eyes closed as though he was trying to shake himself free of a nightmare. "Don't they see—I'm in the car!" he screamed, his face contorted in rage.

***

Jenkins opened the helicopter door. "Damn reporter's more slippery than snot on a glass door handle," he said, reaching into his duffel bag in the back seat. Jenkins picked up a flashlight, a long cord and an Uzi. He stuffed two magazines in his waist.

He fastened the cord to the left helicopter skid. "I'll get 'em down there," he said as he swung the strap of his automatic pistol around his head. "I've got the radio. You'll hear it all." He put on a beret, then removed it. "Where's Frank?"

Bueller turned on his walkie-talkie.

"How far are you?"

"Just got on Postal Route 248. About 5 minutes from da tunnel." Jenkins heard through the walkie-talkie.

"Roger. Over and out." Bueller went to the back of the helicopter and from the custom-fitted gun rack, took a Browning rifle with scope. He peered through the scope. "I got the exit covered, if they try to run."

"Don't worry," Jenkins replied, "ain't been a liberal reporter I know that's survived a shot to the chest and head . . . even one whose brain's are workin' overtime. Aren't these human interest guys usually dumb?"

Bueller shrugged his shoulders. "He's lookin' a lot smarter than you just about now."

Jenkins grinned. "Yeah, and will for another three minutes." He grabbed the cord and rappelled down the hill. As he dropped to within twenty feet of the road, he stopped, unstrapped his unregistered 1978 Uzi and put his right finger through the trigger guard. He dropped, crouching, to the ground, gun at hip level. His heart was pumping harder. Once again, he reveled in the adrenaline rush. The tunnel was dark, reminding him vaguely of the mosques he had entered in Fallujah five years earlier. Only this time, the target was unarmed . . . and he wouldn't be blowing the structure up afterwards. *Piece of cake.*

The car was not in the front forty yards, which was as far as the natural light extended into the unlit tunnel. As he entered the cave, the wind picked up. The howl deafened him.

***

98

Paul saw the outline of a guy, about 250 yards ahead, sweeping from side-to-side with his flashlight. He bent his head farther to the right, to get a better view of the man through the cracks in the windshield. They also gave a crackle pattern to the flashlight beam. Through the holes the bullets and shrapnel had made, he felt the chill of the wind now whipping through the tunnel. A few snowflakes blew past the car.

Come on, you have a plan A and plan B for everything. He had to. Growing up country poor, life hadn't been a clean and hospitable playground for him. Now though, he'd need to draw on his last reserves of country smarts, if he wanted to leave this tunnel alive.

Trying to drive around the guy wouldn't work. He would have to confront. He turned to Kendrey. "Kendrey, stop shaking, we're safe."

"I'm gonna kill that bastard."

"That's the spirit!" Paul said, taking his hands off the wheel and crossing them in front of him. He drummed his left fingers on his right bicep. "Josh, how good was that team you played on?"

"B-league city champs, f- five years back."

"Were you a decent batter?"

"About three-fifty. A few ho- homers a season."

"That's goo-o-d! Remember the feeling, you know, connecting?" he asked as he leaned back and picked the bat up from the back seat.

"Yeah, sure. N-Never forget."

"Okay, then. Here's the plan. If I don't get him with the car, you lean out and whack him."

"I don't—" Kendrey started, his jaw locking again, "know, but, but I'll try."

"Josh – remember what the little green guy said?"

Kendrey looked at him and shrugged his shoulders.

"Try not. Do or do not. There is no try."

Kendrey nodded.

He shoved the bat into Kendrey's hands.

Kendrey smiled.

"Wait. Lean out and take a practice swing." Paul started the car up. It purred inaudibly.

Kendrey rolled down his window. He leaned out and swung the bat.

Paul shook his head. "You didn't tell me it was the women's division."

"I can't get the t-t-torsion. I'll fall out of the car,"

"What if I hold onto you?"

Kendrey nodded.

He grabbed Kendrey's waistband and held it firmly.

Kendrey held the bat high up, a choke hold.

"Take another short swing."

Kendrey took a solid swing. Paul could feel the body turning. It was an awkward motion, but not hopeless.

"That's more like it."

Kendrey sat back down into the car.

"Now, Josh, when I say 'duck', you duck, and fast, okay?"

"Yeah."

"You ready?"

Kendrey nodded.

He pressed down slowly on the accelerator. The car silently gathered speed. Then the gasoline-powered motor kicked in. Within 150 yards, Paul watched the speedometer hit 65 miles per hour. As the wind pushed through the hole in the windshield, the 'whush' of the gusts turned into a high-pitched howl. The draft of air chilled his face and injured finger.

A man with an automatic, eighty yards ahead, was slowly approaching, threatening his quest for the truth, and his life in the bargain.

16

Sweeping his flashlight and gun from side to side, Jenkins continued into the tunnel. The howling of the wind had diminished a little, but it was still the only sound he heard. He was now straddling the lane divider in the center of the tunnel. It was strewn with broken bottles and cigarette butts. *The bastards probably left out the other end. The ground crew will get 'em. Too bad.* He was now sixty yards into the tunnel, and finding nothing. Suddenly, his flashlight panned the Fusion coming towards him, at, he figured seventy yards away.

Damn. Why hadn't I strapped the flashlight to the gun? He pressed the flashlight onto the top of the Uzi, holding it with his left hand, then lifted the barrel, jerked the gun to the right and squeezed off two bursts. From the faint reflection of light off the windshield, Jenkins saw he had missed.

Jenkins swung the light and gun barrel to the left. The car now bore down at him from forty yards away. He dropped to a crouching position and blasted, hitting the hood. *Aim for the top of the steering wheel.* He lifted the Uzi higher, squeezing off five more rounds at the driver's side of the windshield.

***

After the first round of shots missed, Paul knew they'd been lucky. But now, this would be a bullfight. The cracked windshield made visibility difficult, and the reflections of the flashlight scattered in twenty spots in front.

Paul lined the gunner up in the center of the front of the car. *Forty yards away.* The gunner was crouching . . . aiming. Too close to miss.

"Duck. Now!" he yelled.

A volley of bullets hit the windshield, pushing Paul's side of it half a foot into the sedan. But Paul kept his foot on the accelerator. He jerked back up—the car had swerved a little far to the right. He adjusted it back . . . twenty yards . . . 15 . . . accelerator full down.

The gunner had to do something or get run down. He was readying himself to jump the only place he could—to the right of the car, gun ready to fire.

"Kendrey. Bat. Now!"

Paul grabbed Kendrey's waistband with his right hand and leaned back, jerking the car going to the right, as the gunner jumped in that direction.

Kendrey lunged full out, holding the bat almost straight—it would have been a bunt. Paul jerked the steering wheel right.

The gunner aimed the barrel at him.

Kendrey extended the bat five feet out of the car. The gunner didn't even see Kendrey jerking the bat down. Two rounds hit the top of the dashboard. The bat then made full contact with the gunner's head. *Thwack-crunch!* Paul heard, as the blow crushed the radio headset.

"*Idiot, it's me!*" Kendrey screamed as he looked back at the fallen gunner.

"Good Job!" Paul yelled, slamming on the brakes. He punched the stick-shift into reverse and stepped on the gas pedal. He looked on the five-inch screen built into the control panel. The reverse lights flooded thirty yards of the tunnel in red light.

The gunner was bent over, shaking his head. He faced the South tunnel wall.

As the car approached to within ten yards, the gunner turned his head towards it. Surprise registered on the man's face. He crouched, preparing to leap to the right side, and off the screen.

Paul felt a thud, heard a 'crack-tinkle', saw a hand and a lower leg flash on the screen. Once again, he slammed on the brakes. He jumped out of the car and ran to the gunner.

Kendrey followed, holding the baseball bat at the ready.

Judging by how the gunner's body was now sprawled out, with the left thigh at a twenty degree angle from the joint, he figured the top of the Fusion's high back red light had hit the gunner full in the hip. The impact had propelled him ten feet back.

They approached the fallen body. Paul nudged him with his shoe. He moaned. His neck was red, probably sprained, the result of his body having twisted around, like an ice skater's, but without his shoulders leading. Out of the corner of his eye, Paul eyed the Uzi.

"Josh, if he makes a move, smack him."

Paul crouched down next to the gunner. He reached down to his belt, and felt from left to right, finally wiggling out two magazines for the Uzi. He reached farther to the right for gunner's pistol.

The gunner opened his eyes.

Paul felt his left wrist being bent backwards. "Unh!" he gasped. "Hit-"

A knife blade tore through his jacket, through his sweater, cutting his skin.

The bat crashed onto the bridge of the gunner's nose.

"Aargh!," the gunner screamed. The knife blade clinked against the asphalt. He was out, his nose strangely leaning to one side and oozing crimson.

Paul knelt for a few seconds, massaging his wrist, looking at the damage the bat had rendered the gunner's face.

"Thanks, Josh. You . . . you're a star," he finally uttered. *What now? Weapon!* He jumped up, ran to the Uzi and picked it up. He tried changing the cartridge. His hands were shaking too hard. The top of the cartridge kept clanging against the trigger protector, then against the short barrel. He brought his arms down, looked up and took a deep breath. Another.

Paul turned the gun upside down. He slowly brought down the dark gray cartridge magazine to the rectangular port. He pushed it down until he heard the click.

He walked around the rear of the car. His mind's eye noted that the red right rear light cover was cracked, with a two-square-inch piece missing. "Shit. This car's a mess."

"Yeah, but F- Ford -engineered."

Paul couldn't help smiling. "Oh well, let's go."

Once in the car, Paul winced. His view was almost completely obscured by cracks in the windshield.

103

"Could you turn up the heat, it's cold in here," mumbled Kendrey, as he settled into his seat.

"It's up all the way," he said, zipping his coat. "Hang tough, Josh. We've only got about thirty more miles."

"I'll f-f-f-reeze to death before then."

"Ahright, ahright. I'll drive kinda slow." Paul started the car up.

"No, no!" Kendrey yelled. He opened the door and got out.

"What the fu?!" Paul muttered.

Kendrey jumped into the back seat.

"It's not going to be warmer there, Josh."

"I'll see."

"Whatever." Paul then thought about his next move. The pilot would be up top, probably with a weapon, so he would have to exit the tunnel appropriately. He had calculated they were about three miles from the turnoff onto Route 30 that would put them back on the fastest route to the Stoltzfuss house.

Paul jammed on the accelerator. They flew out of the tunnel at over sixty and the car was still accelerating. He jerked the steering wheel slightly to the left. A shot hit the back window. "Mix it up!" he remembered the evasive driving instructor saying. He slammed the brakes, then accelerated again.

***

Antoine looked at the brown trees, the overgrown middle barrier, and felt the rough road. He didn't like it. His experiences in tunnels had not been pleasant. In one of them, he had found his brother, strangled to death. The result of a drug deal gone sour, but still. At least he had found the guy who killed him . . . and hung him by his neatly epilated testicles before slashing his throat.

As the tunnel entrance appeared, a few hundred yards ahead, the radio buzzed. Antoine pushed the button, "Yeah?"

"Target just exited Western end of tunnel. Going about sixty." It was the pilot. "Jenkins out-of-contact. Should I follow the car? Over."

Antoine pondered the situation. If the helicopter pilot went after the target, what could it do? The pilot couldn't shoot and pilot at the same time. "No. Just check on the Texan and wait for us. Over."

***

The helicopter could have caught up with them, but didn't. Apparently, that part of their operation was useless without a gunner, and *that* gunner was out-of-action. He looked over at Kendrey. He was still shivering.

Paul heard some papers rustling.

"Hey- wh-what's this about the Danabold machines, not tested enough with seniors? Sure they were."

"Yeah, what. Now at least two complaints about changing votes. Gotta admit that at the very least, they're confused."

"Alzheimers . . . and what's this about dry runs? Parallel testing?"

"Sure, State Department has to do that. They bring some in, randomly and test them."

"I know, but why, to ch-check my f-flawless work?"

He laughed. "Yes, you're perfectly flawless work."

"Do you know how they parallel test?"

"Not completely yet, no, but I've got an idea. And now that I can focus on this, I'll find out."

Kendrey again started rocking back and forth. "W-Waste of time."

"Listen, Josh, you said you'd tell me now. Why are they after you?"

Kendrey opened his eyes wide and shrugged his shoulders.

"All right, all right. Awesome hit, by-the-way, it saved us."

"I wanna call Mom."

"Fine."

Paul handed it over with the battery to Kendrey. "Just keep it short. Battery's running low."

Kendrey nodded as he took the phone. "Where we going?"

"To Millersburgh."

"What's in Millersburgh?"

105

"We've gotta find the farmer who had a complaint about the machine."

"How d-do you know about that?"

"The press . . . " Paul muttered, reaching down and pulling out the paper he had stuffed earlier in his car door. He gave it to Kendrey. "Page two, bottom."

Kendrey took the paper.

Paul heard the rustle as the paper was opened, then folded.

"Hi, Mom. It's me. We're still having fun . . . Yes, I'm well fed." He pulled another Mallomar out of the bag now at his feet. "We're going to Millersburgh . . . yes I know it's getting dark."

Paul heard him biting down on the cookie. "Some farmer," Kendrey continued, chewing between words, "Stoltzenfusk . . . who said the voting machine," *smack*, "did s-something strange. Probably just t-too old to use the machine . . . Where?   What would you c-care, you don't r-read all those papers . . . Okay, it was in the Altoona Brand."

"Bend," Paul corrected, loudly.

"Bend!" Kendrey parroted.

Paul couldn't help chuckling. After all, Kendrey was really being a sport. Although his life was in danger, he could have acted passively, victimized. Instead, he had just saved Paul's life, as well as his own. It seemed that he could overcome his handicap when the chips were down, as long as he had his link to Mom . . . and his junk food. Paul felt he could use a Mallomar himself just about now.

"Yeah, Mom, I know. I sound t-tired so I should lie down. I'm in the back seat now, so I can do that . . . there could be an accident . . . They happen all the time."

Paul drifted off from listening to the conversation and remembered the first time he'd had a Mallomar. He must have been eight. And they were watching the evening news . . . announced the right way, conscientiously, thoughtfully.

"NO, no—we've been th-through this before—I won't do that! . . . That's Sis's problem. Have to go now. Bye."

***

Jenkins woke up to see the copter landing on the road in front of the tunnel, its projectors basking the entire front of the tunnel in light. Bueller jumped out of the helicopter, leaving the blades turning.

Jenkins lifted his head up. Blood had stopped gushing out of his nose by this time, but he could feel the blood drying on his skin.

"You look like shit," Bueller remarked.

"Yeah. They jumped me. No lights, the wind, whacked me in the head."

"Yeah, should never of sent an Army sniper to do a Seal's job," Bueller said, taking Jenkins' left arm to help him up.

"Shuck you," Jenkins retorted, using only the right side of his mouth. "Shit. Can't stand uw." He fell back down, wincing.

"How's your other leg?"

"Right side's . . ." he hesitated, as he shifted his weight on it, "Okay."

***

Antoine glanced into the back seat. John, in the left back seat, screwed the silencer into the barrel of the machine pistol, as he whistled the tune to "Psychochicken." Tom, in the right back seat, opened the breech of the sawed-off shotgun to check that each barrel held a low-recoil shell of tri-ball pellets. The Sedan pulled into the tunnel.

Jameson, between the two, held a grenade in his hand. The driver initially slowed down to forty miles an hour, as he maneuvered around piles of rubbish. Antoine surveyed the graffiti on the walls, written over the course of the last thirty years, by bicyclists, lovers, and, it seemed, anyone not too lazy to bring a can of spray paint or crayons to this mile-long tunnel. He felt the nape of his neck rising as the car proceeded through the cave-like space. He would've liked to have Girelli here as well, but at least he'd agreed the other car, which doubled back through Harrisburg, would pick him up on the way here. A minute later, Antoine saw the end of the tunnel, and the profiles of two people near it. He breathed easier.

107

Behind them, a Bell 5-seat chopper. The black Explorer pulled up to the pilot, now supporting Jenkins, as the latter leaned on the back of the chopper.

Antoine rolled down the window.

"What happened?"

"He said the car backed into him; his hip joint's out or cracked. He's not in great shape," Bueller responded.

Antoine looked more closely at Jenkins. He shook his head. Jenkins' left eye was bright red and swelling; he'd have a huge black eye and Antoine estimated the discoloration would extend to his entire nose and two inches up his forehead. But his right side looked untouched. "Think ya can still shoot?"

Jenkins nodded slowly. "Just can't walk."

"You're tough, kid, I'll give you that much," Antoine added, getting out of the car and closing the door. His cell phone rang. He picked it up and walked slowly towards the tunnel exit.

Jenkins suddenly stopped. "My Uzi . . . It was right there," he said to Bueller, gesturing with his right hand towards the spot he had fallen.

Antoine came walking back to the car. He stuck his head through the car window. "Guys, get out and look for an Uzi." He turned to the pilot. "Bueller, listen up. We just got new info, so the plan's changed. Jameson is gonna go wit' you."

Antoine explained the new directive in thirty seconds.

"And then back to da base. You got gas for that?"

"Yeah," replied Bueller, "but not more."

"Boss, we can't find anything," Tom yelled from forty feet away.

"Okay, let's go!" Antoine barked. "The reporter's got an Uzi and a pistol, but it still don't make that pussy no Clint Eastwood."

17

Paul thought he saw a car behind him, but as the road curved to the left again, he didn't see it anymore. Paul turned around and glanced at the automatic, lying under Kendrey's legs. *Just in case . . .* He reached back and slipped the weapon out from Kendrey's legs.

He hadn't tested this one; it hadn't been the weapon of choice for almost twenty years, well before his gun range had offered any automatics.

***

Antoine knew that the reporter would spot them any minute now, if he hadn't already. And since the road was officially closed to traffic, he would know they weren't here to party. *Given that he's armed, better shock and awe.* "Ready, boys?" Antoine asked as he glanced back at his two men, each a virtuosi with their respective weapons.

"Yeah, boss," John answered.

"Locked and loaded," Tom confirmed, lifting up the shotgun.

"Good. Remember, just the driver. Luke, turn on the high beams," Antoine ordered, as the road straightened out.

The Explorer approached to within fifty yards of the Fusion, the driver turned on the high beams.

***

Seeing no flashing lights, Paul knew the muscular black vehicle meant trouble. *In what form?* He eased off the accelerator. The other car was faster; he would have to be maneuverable.

He glanced to the left, and saw the windows of the black Explorer were tinted—the front ones lightly, back ones completely. He noted the outline of the guy in the passenger seat. He had seen him somewhere before . . . just recently.

The Fusion, now almost coasting, had slowed down from 50 to 40. Paul edged it slightly to the right, but he still had enough space to maneuver.

The Explorer's windows rolled down on his side. *Crazy, in this weather.*

He saw a machine pistol barrel with silencer emerge from the front window and a shotgun barrel from the back one. A man's torso popped up above the roof on the far side of the sedan, holding something black and long.

Paul slammed on the brakes.

The weapons roared simultaneously. A deafening *Whoosh* roar. The front linear foot of side window obliterated. An explosive wind. Pain in his knuckles, on his chin. *WHATTHEFU--!* His left glove erupted with a spurt of blood as a shower of glass followed.

"ARRGH! What the hell!!" screamed Kendrey from the back.

As the brake pads dug into the discs, the Fusion slid about thirty feet, while the sedan continued ahead, now also in a full skid. Paul watched as the thin coating of snow was kicked up by the SUV's locked tires. When both stopped sliding, the Fusion was twenty yards behind the Explorer.

He switched gears to reverse. The bottom half of the rear windshield resembled a crackle glass lampshade, scrunched inwards in parts.

He glanced back at Kendrey; he'd been hit in the left forearm too. Paul jammed the accelerator pedal to the floor. His adrenaline surged again when he saw how quickly the Explorer was backing up, spitting snow from its tires.

"Ow this hurts!" wailed Kendrey.

He ignored Kendrey and barely noticed that his own left ring finger was dangling loosely over the steering wheel, the tendons severed from the bone. With his right hand he grabbed the Uzi from his lap and shoved the barrel through the huge gap the shotgun blast had made in the front of his window. The barrel was short. Wouldn't be visible from the Explorer. *They don't know what they've got coming.*

As the Explorer backed up towards them, he accelerated the Fusion forward. He aimed the Uzi's barrel at the rear right tire of

110

the fast-approaching vehicle, its reverse gear lights still burning bright red against the gray backdrop.

The Explorer's driver slammed on the brakes and accelerated forwards.

But the Fusion was now traveling twenty, while the Explorer had just hit five miles an hour.

As his car approached the slower-moving vehicle, Paul pulled the trigger. The bullets ripped the Explorer's back right tire to shreds. But with his glove on, he didn't have sense of touch. He'd pulled too stiffly. *Too long. Too many bullets.* The Uzi kicked up, shooting into the passenger compartment and into the sky.

With the tire burst, the Explorer fishtailed right suddenly, sharply. The front passenger emptied the magazine. A spray of bullets peppered the windshield of the Fusion and in front, missing Paul widely.

The scar-faced man leaned out the far side back seat window and shot over the sedan into the Fusion. 'Clink-crash' Paul heard as the projectiles obliterated back seat left window. *No shotgun blast this time.*

"Aargh!" screamed Kendrey.

Paul heard a 'clank' and the Fusion jerked to the left. *The Sedan's front fender's nicked my back. Shit!*

His foot was still full on the accelerator, so the Fusion started spinning. The trees, the barriers, the mountain, the Explorer swirled around at least three times, peppered with snowflakes. When the spinning finished, Paul was facing back towards the tunnel. To his left, sixty feet behind him, the Explorer coasted ahead, the front facing into the shoulder, away from the center divider, back right tire flopping left and right.

Through his rearview mirror, Paul saw the two thugs in black crouching. "Stay down Kendrey!" Within a second he had angled the steering wheel towards the center of the gently bending road, and ducked down.

Three shots pierced the back window and Paul heard them sail above through the useless drape of glass that was once a front windshield.

He jerked his body up, in time to straighten the Fusion, which had been heading into the trees to the left. He ducked again.

111

One more round sailed through, then nothing. Jamming the accelerator, Paul heard no more 'tat-tat-tat's.'

They were out-of-range.

Paul sat up. The barrel of the Uzi burned his right thigh. He chucked the weapon into the passenger's seat. "*Holey* fucking *moly*!"

The Fusion now barreled calmly down the gently curving road. But his heart pumped uncontrollably while the wind chilled his sweat to near-freezing, his eyes alert to every snowflake on every branch of the pine trees whizzing by.

"You okay?"

"My left hand. It's gonna fall off," Kendrey replied, defeated.

Basically unscathed. But before he could savor the miraculous escape any longer, he had to get back on track. He slowed the car down. *What now?*

He glanced to the left and to the right. To the left about two-hundred yards up the road, he saw a gravel rest area. It had been abandoned for thirty years, but it was still flat. As the Fusion got closer, Paul saw some broken bottles on it. "Yes!" When he didn't come here with his cousins from his Uncle's house, he usually came up a narrow dirt access road on this side of the tunnel. The dirt road would intersect with a gravel road through the State Park and onto 915 or back to route 30.

He looked in the rear view mirror, just in case. The Explorer was nowhere in sight.

***

Removing the spare tire from the trunk in seconds, Luke quickly set about changing the flat, with Parkett's help.

Antoine, in the meantime, held his ungloved right index and middle fingers to Tom's jugular in the back right seat. Blood oozed over Antoine's middle finger and dripped onto the seat. Tom stared blankly at the rear-view mirror of the car. He coughed up blood. Within ten seconds, his entire lower lip was coated a deep crimson. His face was pale.

112

"You'll be okay, Tom," Antoine said, as he pulled a handkerchief from his jacket and dabbed away the blood. He was now down two men. "Remember dat' night at Jocanthes?"

Tom didn't respond. He kept staring forward. He coughed once more.

Parkett looked up from the wheel, now mounted onto the screws protruding from the metal plate. "How's Tom?"

Antoine looked Parkett square in the eye. "He's doin' great for someone-about-to-meet-his-maker. Just shut-up and change the fuckin' tire, will ya'?!"

Parkett, even though a full six foot-three-inches, shrank back. He returned his gaze to the wheel.

"Where the fuck did that reporter bitch learn to fire an Uzi?!" screamed Antoine. "And why didn't anyone tell me he could shoot?" He pulled the lifeless hit man's eyelids down. "Parkett, Luke, put him in the back." Antoine pulled out the two-way radio. "That journalist has the luck of the Irish . . . but he'll get his due at the end."

"Helicopter one. Reading," Bueller's voice responded.

"Target got away. Need ya' to turn back and locate him, over!" barked Antoine.

"Roger."

***

The metal gate, painted a bright yellow, was a standard swing gate, fastened at the end with a short chain and lock.

Paul pulled out the Uzi. Two short bursts and the lock was open. He took the chain off and pulled the gate open. He got back in the car and sped off. He didn't even bother closing the gate behind him.

He drove a hundred yards past the entrance, and the road narrowed. He slowed to 20 miles per hour. In the rear-view mirror, he noticed some black smoke dispersing in the air behind him. Then he heard the distinctive 'thwack-thwack-thwack' of helicopter blade.

***

113

"What does the gravel road feed into?" Antoine asked.

"Nine-fifteen to the North . . . and thirty to the South." The response came, crackling, on the radio.

Antoine glanced at Luke, who had just finished tightening the last bolt. "Gravel road connecting nine-fifteen and Route thirty?" Antoine asked.

Luke nodded.

"Should we stick with him to find out which road he's taking?"

"Naw don't bother, Bueller. We'll find him either way, it's snowing. Report soon."

"Roger. Over and out."

Antoine shut the radio off. He turned to Luke. "It looks like we got some good news after all. The guy's engine's smokin'."

The three piled into the Explorer. Luke turned the sedan around and accelerated forward. Antoine noticed a bullet hole had pierced the top of his door window. He turned the heat up and adjusted the vent direction toward the back.

***

Smoke had been bellowing from the exhaust pipe for several seconds, but Paul had been oblivious to it in the diminishing daylight, as gray cumulus clouds began obscuring the last of the twilight. But now, he couldn't ignore the wisps emanating from under the hood. Bad engine trouble. Shit. That explained the jerkiness.

"We're in trouble, Josh. I think we're gonna have to borrow another car soon."

Three hundred yards more of fishtailing and they reached the second gate. Paul stopped and jumped out of the car, leaving it running. He opened the trunk, pulled a combination lock out of his gym bag and stuffed it into his coat pocket.

Once again, he shot the cheap lock on the chain wrapped around the barrier gate. He pulled the lock off and tossed it in the bushes. He jerked the chain off. It fell, clinking against itself, into the snow. After driving through, he got out, closed the gate, and retied the chain through the loops.

He thought he heard, between gusts of wind, the sound of a car driving some thirty feet above him.

Kendrey got out of the back and sat in the front passenger's seat.

Paul attached his own lock onto the chain and turned the tumbler. He got back into the car. "Now let's see, Josh, if that line about Master Locks' shot-through-and-through but-still-working is true. It'd better be."

He drove down the 20-degree incline for forty yards. Once at the bottom, he turned left, in the direction of Route 30. *Don't know how long it'll throw 'em off, but it's worth a try.* He drove fifty yards in the thinly snowed-over gravel path, then slowed down, and drove the right side of the car into the ditch. He stopped and drove backwards, his left wheels hugging the right edge of the road, while his right ones continued driving in the shallow ditch. After driving backwards the length of a football field, he noticed the road widening and executed a K-turn.

As they proceeded north along the road, they passed the abandoned ranger's cabin and the road again narrowed.

All the while, Kendrey rocked back and forth, holding his left forearm with his right hand.

He had the eyes of a wounded dog, unsure as to what had happened, or to what lay ahead.

A minute later, Paul saw the T-intersection marking the end of the gravel road. Route 915.

A beat-up Chevy passed the intersection, forty yards ahead, on its way to Route 30.

Damn, missed that one. As the Fusion reached the intersection, he looked both ways. No lights, nothing, from either direction. He turned to Kendrey. "How're you feeling, champ?"

"How do you think? My arm is f-falling off!"

"Show it to me."

Kendrey slowly moved his hand, caked with dried blood, towards him.

"Can you move your fingers?"

Kendrey flexed his fingers. "Ow, it hurts." His eyes welled up with tears.

115

"Josh, it may be full of shot and glass, but from the look of it, it's nothing deep, I mean there are no dangling tendons," he said, pulling his own left hand in front of Kendrey, "like this."

Kendrey looked at Paul's ring finger, the tendons bulbous under the dark red coating, and nodded.

"Tonight I'll take you to the hospital," Paul mumbled.

"I'm still—" Kendrey continued, his jaw locking for two seconds, "still f-freezing, too!"

"So am I, bud, so am I," Paul responded. He wondered if it would be before the Explorer showed up. He looked back through his left cracked window towards the gravel road they'd just left. He saw nothing through the trees. "We'll just pay a quick visit to the farmer, and be off. He's waiting for us, with hot coffee."

"Farmer who?"

"Stoltzfuss – he's the farmer in that article."

"Oh yeah."

Paul felt his left hand throbbing. The blood had congealed in a brown mass on the middle of his glove, making it almost indiscernible from the glove itself. He started calculating how long it would take to change the tire on the Explorer.

"If it takes you longer'n four minutes to change a tire, go put on a dress afterwards," his dad used to say, in the age before women served in the infantry. Paul smiled briefly as he imagined the Scarface in dress. His smile disappeared as he reflected on the sobering fact— three minutes, before the Explorer would be on the road again. Thirty seconds on the first gate. If Master Lock held out, it would buy him another three minutes at the second gate. The tracks in the wrong direction . . . another minute-and-a-half, not more. It gave them eight minutes, max.

18

The engine felt heavy and unresponsive. Paul heard a mini-sputter. No giving out now, no way. He gunned the engine, in hopes that the heater would push more warm air through the vents. Nothing. However, the smoke tails rising from either side of the hood slowly made their way, through the porous windshield to the front seats.

Kendrey coughed.

Suddenly, something flashed off the crackled glass windshield.

Paul turned to his left and saw headlights in the distance, moving southward along route 915. They were steadily making their way towards this spot. He looked ahead and behind him. Anybody seeking to go around either end of the car would put their car in a ditch.

He once again looked left, through the gaping hole in his window. The car was about half a mile away. From what he could see in the diminishing gray light, it appeared shiny and dark.

Without turning away, he cleared his throat. "Josh, could you get out and wave that car down when it approaches?"

Kendrey nodded.

Paul got out, stepped to the back and lifted the right passenger door handle. There was no resistance, but the door lock didn't budge. He saw a small hole next to the handle. *Buckshot or tri-balls.*

He reached through the blank space where the window used to be and picked up the Uzi. He took the used magazine out and threw it in the back seat. He reached into his coat pocket, pulled out the last fresh magazine. It felt cold, solid. He clicked it into place and turned to face the oncoming car. It was now two hundred yards away.

Paul moved the Uzi behind his coat with his right hand. He had thought quickly, if not terribly hard, about this move.

The big dark car approached to within a hundred yards.

From his left, between the frigid gusts, Paul heard some staccato knocks. It could have been a woodpecker. But he well knew woodpeckers didn't peck this late. The hairs on the nape of his neck stood up.

The large black car slowed down.

Normally, he would identify himself and wait for someone to recognize him, at which point cheery banter would follow and within a few minutes the stranger would cooperate. But today, an Explorer could very well pull up and its passengers unleash dozens of rounds into their cheerily bantering bodies. No, today would not be 'normal.' He also knew that with Durand dead on the apartment floor, with his own prints everywhere and now with a carjacking, he had no choice but to pursue this thing to its end. He would explain to the police later . . . if he could stay alive to do so.

"I'm freezing!" Kendrey yelled.

"Just wait a minute!"

"I c-c-c'ant!"

"Shit!"

"Don't get mad, Robinson. I have needs!"

As the car approached to within forty yards, Paul could see from the form that it was a hefty BMW. The car's high beams flashed.

Paul waved his left arm.

He saw, out of the corner of his eye, Kendrey doing the same.

The BMW 740 slowed down. The driver, with mobile phone clasped to his right ear, rolled down the window and flipped a birdie. The black vehicle nonetheless came to a complete stop, twenty-five feet away.

An impeccably dressed man, Paul guessed to be in his late twenties, with short jet-black hair, jumped out of the car. He shoved his cellphone into his pocket. Like a bull, he charged towards Paul, stopping twelve feet away.

He had broad shoulders, and, Paul could surmise, a well-developed torso. If it came to a brawl, he wasn't sure this guy couldn't pummel him into the ground.

"We had some car trouble—" Paul started, friendly enough.

"Your problem, asshole. You have three seconds to get back in your car and move it, before I flatten your face. One . . . two . . . " the driver stepped another three feet menacingly towards him, his face contorted in rage.

"Okay!" Paul stepped back. *The guy doesn't recognize me.* Today, that was a good thing. "Hey, you sound just like the owner of a black luxury vehicle!" he said jovially. "You know, I've always wanted to do this."

The guy took one more step towards Paul, his chest out and his hands balled up in fists.

Paul whipped out the Uzi, and aimed it at the man's chest.

The driver jerked his head back. He froze in place.

Without looking away from the man's eyes, Paul lowered the barrel and gently touched the trigger. Snow mixed with asphalt leaped up in all directions in front of the man.

The driver hopped backwards from foot to foot, thrusting his arms into the sky. "Shit!! Wait, WAIT!" His arms started shaking. "Okay, Okay! Whaddya want?"

"I need your car and your phone."

"Fuck you. Goddamn hillbillies."

Paul lowered the gun to his waist, and squeezed off a round between the driver's legs, keeping his left hand firmly on the top of the stock to prevent the gun from kicking up and damaging the man's family tree, irreversibly. A deep ache racked his left hand.

The driver turned around and lifted his expensive-looking coat tails to his face. Light from the BMW beams showed through a hole. "It's cashmere, you asshole . . . dja even know what that is?!"

"Do we have a trade?" Paul countered. He brought the Uzi up to his right eye and peered down the sight, aiming at the middle of the driver's forehead.

"Yeah, *yeah*! J-j-just be cool," the driver relented, shaking his head like a dashboard doggie.

"Good. Now throw your phone to your left."

"Hey man, it's an HTC – I just shelled out five hundred bucks . . . Shit, all those fuckin' apps."

"Didn't you back 'em up in your computer, or the cloud?" Paul asked.

The man shook his head.

"Hillbilly." Paul jerked his head sideways. "Josh, come here." He turned back to the driver, keeping the gun squarely aimed at the man's chest.

119

"Kendrey, keep this sociopath under control," he said, handing the Uzi to Kendrey. "I'm gonna get the Fusion out of the way."

"Do a farm foreclosure and get car-jacked by rednecks as a result," the lawyer muttered.

Paul shrugged his shoulders, scooped the HTC off the ground and got in the car. He turned to Kendrey. "Oh, and if he moves, kill him, just like in Quake," Paul said, loudly enough for the man to hear him.

To his left, he heard a loud 'crack' through the wind, from the direction they had come. It was more than just a handgun.

Paul backed the Fusion to the spot he had picked out earlier, about thirty yards from the T-intersection, midway into a hundred foot stretch of road that inclined upward to meet the paved route 915. He swung the Fusion around, almost backing into a tree, before effectively blocking the road with the car, a hundred feet from 915. The smoky air pinched his nostrils. He got out of the Fusion and looked around.

Getting back into the quasi-wreck of a car, he coughed. He edged it a foot back and pulled the parking brake as far back as he could.

Any minute now, they'd be showing up. But still, he had to do this. Paul took the Leatherman Wave out of his left coat pocket and pulled out the short sharp blade. He got out and perforated the bottom of the front left tire with two quick jabs. He sprinted around the car to the other side. He repeated his perforation with the left and right back tires.

From the back of his car, he removed seven of the longest pitons, and his Bongo climbing hammer. *Finally, a good alternative use for this gear.*

All he needed was a delay. He went to the left front and slid under the car. The snow made it easy for him. He jammed one piton between the brake disc and the brake pad. He slid back out and crouched down, taking a piton from his pocket. He drove it through the tire into the ground, repeating the same in the back.

He stood up kicked snowy gravel onto the pitons, as he listened to Kendrey singing "I Want Money." Paul glanced up at the scene.

Kendrey was gyrating his hips, but still keeping the gun aimed at the driver. Now, that driver looked terrified.

That programmer is nuts. Paul completed covering the metal loop head with the gravel, dirt and snow. Sweat slathered his entire back. After repeating the same on the right side, he picked up his backpack and slung it over his shoulder.

He ran back up the slope. As he stepped onto the asphalt, two drops of perspiration fell off his nose,

Kendrey glanced at him. "Did you take the snacks?"

"What, like a pastrami and swiss on rye?"

"That sounds good."

"Kendrey, I'm kidding. Are you nuts?!"

Kendrey's expression turned dog-like again, with a glint of something else. Paul thought he saw his jaw muscles tighten in the waning light.

19

The Explorer stood in front of the lower gate, at the end of the access road. Antoine fired his Glock again. The bullet ricocheted off the lock, leaving a slight dent. Parkett nodded. Antoine pulled on the lock. It held. "Goddamn it, get me the axe." Parkett reached into the back seat and gave it to his boss. Antoine made contact with the chain and the axe glanced off one of the links, scraping a portion of the paint off. "Parkett, hold it straight for me."

Parkett bent down, took both ends of the chain and held them taut. As Antoine hit the chain, the ax head bounced back violently, narrowly-missing Parkett's forehead. After another seven hits though, the chain snapped and Antoine threw it to the side.

The Explorer drove through and Luke saw the road, covered with a thin layer of snow.

Antoine looked ahead. The tracks of the Fusion were clearly visible; they were dark on the white snow. The Explorer reached the bottom of the incline, some forty yards below. The tracks headed to the left. They were going for Route 30. Antoine pointed with his jaw to the left.

After the Explorer turned left, Antoine focused on the road. After a minute, he focused into the distance, hoping to see red lamps. He carried on for a couple of minutes, before returning his gaze to the road just ahead. No tracks. "Stop, Luke."

Antoine stepped out. He walked a couple of yards in front and bent down, first to the left, then to the right. He lifted his right hand, spun it around, and pointed to the back of the car. He returned to his seat. "He musta' back-tracked."

***

Paul ran back to the gravel road and fifty feet down the incline. He jiggled the back right door.

It was locked.

He thought he heard a car in the distance. But the wind continued howling through the trees, so he wasn't sure. He ran to the other side of the car, reached in and pulled out the two grocery bags.

Suddenly, the car interior was flooded by light coming from a fast approaching car.

Will I make it? He turned around to see the two thugs, already out of the Explorer and taking aim at him.

Holding the bags, Paul ran to the other side of the car. He dropped to his knees. Shots peppered the doors on the opposite side of the car and more glass tinkled. The car shook with every shot.

Goddamn junk food. If it doesn't kill you one way, it kills you another. How long would it take them to cross a hundred yards? Nothing flat.

Without looking back, Paul whipped out his car keys and pried open the gas tank cover.

"Give it up and no one gets hurt!" Paul heard in the distance.

He ignored more shouts as he unscrewed the gas tank lid. He emptied one of the bags and vigorously rolled it on his lap it until it vaguely resembled an 18-inch long brown joint.

"Throw the gun to the side, and we won't shoot," he heard from, he now estimated, forty yards.

At least they're not reckless, he thought to himself as he stuck the brown hoagie into the tank.

He pulled out his plated Elvis Zippo lighter, and lit the bag. He then pulled out the shiny aluminum foil-wrapped Mallomars from the pile of items. He ripped off the end. He held them like a gun to his chest. He got up, standing straight.

The shots whizzed by. One, he felt perforated his jacket on the left side, missing his arm by half-an-inch. He crouched down again. But he had succeeded. They had stopped moving forward. The big square-jawed guy even ducked.

Paul picked up two of the chips bags, threw them in the other bag and, still crouching, hobbled ten feet towards the incline. He looked back and saw the bright orange flame disappear into the tank. *Now.*

He ran up and to the left for two steps. He promptly switched to the right and two shots whizzed by. Suddenly, the night changed to orange. The trees turned a surreal greenish-orange, while an other-worldly force lifted him the last thirty feet to the road. Paul sensed the back of his head burning. He fell on all fours on the road. *Ow, it burns.*

He jumped up, patted the back of his head, putting out some embers. *No time.* He heard footsteps next to him.

"Awesome, Robinson! But that'll reduce its trade-in value, you know," Kendrey yelled.

"Wise ass!" Paul hollered back, "Let's get outta here."

He glanced backwards. Nothing was visible. Not pursuers, not a car. Just a large orange ball.

He held the door open for the driver.

"Come on, we'll drive you to the next town. You'll hang out there for four hours and I'll get the car back to you."

"I'll take my chances with them, asshole."

"Chances are, they'll kill you."

The driver opened his stance and crossed his arms.

"I'm not bullshitting you. Come on," he repeated.

The BMW-driver cocked his head and flipped Paul a salacious birdie.

Paul grabbed the Uzi from Josh and pointed at the lawyer. "Get in, Josh. . . and you, Mr. Hotshot, now."

The BMW-driver kept his arms folded. He stood up straight.

"Come on, come on," Paul said, opening the passenger door with his free hand.

The lawyer turned on his heels and ran towards the burning car. He kept extending and then retracting his arm, his middle finger protruding skyward, lit by the dying flames of the Fusion.

Paul shook his head, bumped the passenger door shut with his hip and jumped into the driver's seat. He leaned out the window. "Back in four hours." he shouted, tossing a bag of spicy tortilla chips into the road.

As he completed the K-turn necessary to put the BMW back on the northwest-facing lane, he glanced to his left.

124

One of the thugs had made his way through the trees and was about thirty yards away, his profile well-lit by the flames. He stopped and lifted his pistol.

Paul jammed the accelerator. The car flew out. *Wow, this baby flies!*

Kendrey stuffed a fistful of chips into his mouth, his eyes rolling wildly.

"I wonder how much this thing set him back." Paul continued. In the distance, now some hundred yards back, he saw the thug come out onto the road, holding the gun.

"Damn, Josh, I should have fired some rounds, forced him into the car. But we just didn't have the time." *Maybe they'll find a common language.*

"Could be. S-stranger things have happened."

***

Fifteen miles away, the copter flew just above the trees.

The left side of his head was throbbing. Jenkins had taken the painkillers, but they took only the sharpest pain away. From what he recalled his buddies saying, the pain would become intolerable at around midnight.

It was dusk, but the Bushnell Elite rifle scope provided enough light for him to see the buck clearly. Jenkins took aim. He squeezed the trigger. The shot kicked his shoulder back. The kick spread to his head and he felt pain in his nose.

The large buck continued weaving between the small grove of trees, then ran into an open space. The buck accelerated elegantly through the snow. Jenkins fired again. The beast faltered, but continued running, a little slower, then slower still. He jerked to the left, then to the right. His right back leg stopped moving. The buck collapsed on its side.

"Good shot," Jameson piped up from the back.

Jenkins said nothing. The copter hovered over the animal, as Jenkins fired another .308 round. *I will at least be partially redeemed,* he thought, as the copter landed.

***

125

Antoine sized up the situation instantly; so this is the sucker whose car they'd stolen. Before the guy could say anything, Antoine yelled, over the blare of the alarm:

"Kid, we'll give you a ride to the nearest town. Just help us push it out of the way!"

The BMW-driver nodded. The three got behind it and pushed. There was a creaking sound, but the car wouldn't budge.

Their glances met and almost simultaneously, all three took their hands off the piping-hot steel. Their gloves were all smoking.

"What the hell?" Antoine yelled. He turned around. "Luke, need ya help."

The driver jumped out of the sedan and joined them. It still didn't budge.

"Let's swing da back out," Antoine suggested.

Once again, it moved nowhere.

"He must have spiked the tires or jammed the brake pads. We'll have to roll it," shouted Luke.

The four lined up on the driver's side of the car and started lifting up. The Fusion rose two inches and stopped going up. *I'm too old for this shit.* They let it drop.

"Name's Phil Crossman, by-the-way," the ex-BMW driver said, offering a gloved hand to shake.

Antoine didn't answer.

"And you?"

"That asshole just stole three million from our company, the fraudster," Antoine replied.

"I heard him hammering stuff, metal to metal. Sounded like he was working the rims," Crossman said.

"Dja hear that Art, Luke? The rims!" Antoine shouted to his cohorts.

Luke got out of the car, and went to the front tires. He dropped down on his knees, next to the smoking tire.

"He's jammed some wedges here into the disc brake," said Luke.

"Ouch, he's hammered the back tires into the ground," yelled Parkett a second later, shaking his hands as he tried lifting the pitons out with his steaming gloved hands. Apparently, they didn't budge, as Parkett's grunts turned louder.

126

"I'll get the crowbar," Parkett added, sauntering to the trunk of the sedan.

Antoine watched, thinking how slow Parkett was moving.

"Take the lead out. Let's move it!" he finally shouted.

The BMW driver turned to Antoine. "You guys saved my bacon. If you need any free legal counsel—"

"So, you're a lawyer." Antoine smiled.

"Yeah, and the guy was traveling with someone," he interjected, "A Kendrey, someone. Josh Kendrey."

Antoine stopped smiling. His face turned serious. He looked down at the ground.

"So what kinda car ya got?"

"A BMW 740."

"Color?"

"Black…license plate S-E-X-Y-0-1-0".

"Hmm, a pussy magnet," Antoine interjected, getting angrier and angrier with every second, angry that the target was getting farther and farther away; pissed that he was going to have to do what he was going to do. And anxious now that all hopes would soon rest with the helicopter crew, the one that had done such a bang-up job in the first place.

Antoine turned around towards Luke, who had now pried one of the pitons out of the ground with a hammer. "Hey Luke, could ya jest bulldoze 'dis thing outta da way? We lost ten minutes here already."

"I dunno boss. The motor's in the front, but it's still a solid car, and I'll be pushin' uphill. It'll only be another minute on this side."

"Done!" yelled Parkett, as he held up two pitons.

Twenty seconds later, the four of them stood behind the Fusion. They lifted it up and moved it a foot to the left before putting it down. They lifted again, moving it eight more inches.

After the third lift, Luke said he had enough space.

Luke walked back to the Explorer.

Antoine followed Crossman's eyes as he focused on the gunshots in the driver's window of the sedan.

Antoine pulled out his Beretta PX4 Storm with silencer and held it behind him. "Have you ever seen such a sunset guys?!"

Antoine said, as he looked past Crossman, beyond the trees to the West.

Crossman turned to look. "What suns--?'" were his last words, and a soft "Phutt" was the last thing he heard, as he fell to the ground, his brain and skull now two ounces lighter.

Antoine looked down, shook his head and re-holstered his gun. "Now drive, Luke, like the fuckin' wind on fire."

.

20

Latanya loved teaching for the 'Aha' moments she read in the eyes of the more astute students. The class she was lecturing for, "The Social Contract and Politics," a requirement for poli-sci majors, involved complex discussions of people's comprehension of that contract.

Her mid-point in the semester was John Rawls. "Polls conducted have demonstrated people's understanding of the inherent fairness of America's original founders." She paused. After all, what could be fairer than not knowing whether you will enter life and this society blind, mentally-retarded, dark-skinned or blonde, autistic or athletic, into poverty or wealth, handsome, beautiful or coyote-ugly, female or male, brilliant or below average." She waited for her students to visualize the spectrum. "John Rawls' philosophy—which describes America of the 1960s, doesn't argue to compensate for the deficiencies using some well-defined scale, but strives to assure that the benefits accrued by the athletic, handsome, white, male, and intelligent, also benefit the less-advantaged."

A curly-haired student thrust up his arm.

"Yes, Andrew?"

"But what if the person achieved the benefits, let's say a big salary, through years of hard work. Would that be fair to him?"

"Good Question. Rawls would say that even that man had been blessed with the propensity to work harder than others. Note that Rawls didn't say that the hard-working person shouldn't benefit disproportionately from his work, only that he shouldn't benefit to the *exclusion* of the less fortunate."

"Sounds like socialism to me," Andrew boomed.

Three others clapped. "Yeah," they each repeated, almost in unison.

"Are you the oldest son of the family?"

"Yeah, and so what?" Andrew asked.

"Could I ask that all the eldest children in their families raise their hands?"

About two-thirds of the students raised their hands.

"A majority of college students are the oldest children of their families. For a number of reasons—emulation of parents, sense of responsibility, and, most importantly, unfettered attention during the first one-and-a-half to two years of life—made you leaders. But what did *you* do to get that first spot in your mother's womb?"

The lecture hall fell silent, but Latanya could see the 'Aha' gleams in the students' eyes. Andrew was shifting nervously in his seat. That's the way it was with so many eldest children —a sense of entitlement brought on by enormous egos.

"And today, we'll try to understand why the attitudes of the four major rural communities of Pennsylvania fundamentally differ with regards to this basic precept that described the U.S. moral landscape for forty-five of the last sixty years."

She went on to describe the Amish, the small Scottish-Irish farmers, the laborers, and the mine workers and what they derived from their participation in, or exclusion from, the Social Contract. But her mind wandered for a split second to her conversation she'd had with David after the shooting range; *would they understand the consequences of the old, malfunctioning, voting machines.* Would they sacrifice their voice with a whimper and concede to "fate," or would they storm the halls of government, as patriots once stormed the ships of the British East India Company, to protest the denial of their right to have their vote counted?

Her thoughts then drifted to Paul. She would have to make a decision. She took in a deep breath. *In front of a punching bag, where I've made all my best life decisions.*

***

His jaw and nose still throbbing, Jenkins nonetheless took in the fallen carcass, the wildly yelping sheltie running between the copter and his master's body and sighed deeply. He lifted his gaze past the grove of trees, which were now becoming outlines against the dramatic cumulus clouds in the distance. It all struck him as being "melancholic"—the word his crazy mother was so fond of, but in its own way, beautiful. All of it a worthy sacrifice for the future America. The thought gave him a warm, fuzzy feeling.

130

Paul turned right on the access road that would, according to the Google map he had vaguely committed to memory, lead to two houses. He drove quickly, and swerved to the left onto the fifty yard mud and gravel path that led to the house.

He skidded to a stop, walked to the front door and rang. No answer. He waited ten seconds and rang again. Again, no answer. Paul turned the doorknob and opened the door.

An eerie silence reigned. The fireplace had a log, still burning a little. A cat approached and wrapped itself around his legs. It meowed.

"Hello! Mr. Stoltzfuss?!"

He crouched down to pet the cat.

"There's something odd here, isn't there, Josh?"

"Like, w-w-why isn't someone here, watching the home fires burning?"

"Right. Or the dogs. A doghouse, but there are no dogs barking."

"Yeah."

"Josh, go ahead and clean your arm. I'll bandage it." Paul looked on the table and spied a note: *Honey don't forget, prayer meeting at 5:00 at the Nielmann's. I'll be back by 4:30. Erna.* He continued looking around the room. The gun cabinet door was ajar. Paul stepped towards it. A rifle was missing. "Maybe Stoltzfuss's just making his rounds of the hacienda."

Kendrey said nothing, as he turned on the water and put his bloody left hand under it.

Paul returned to the kitchen, opened the cupboards, until he found the one with first aid supplies. As Paul removed his right glove, a sharp pain shot through his left hand. The middle finger felt swollen. But he didn't want to deal with his left hand now, so he kept his left glove on. *There's no way they could have arrived before us. I took the shortest route, most of it at seventy miles an hour.* He removed the isopropyl alcohol, gauze and tape from the corner cabinet. As he patted Kendrey's forearm dry, he saw the buckshot pellet. "You know, Josh, you got lucky. This shot probably busted the

131

window and ricocheted off something else. Normally, it would've cracked a few bones and settled in deeper."

"I should feel . . . ouch . . . " Kendrey reacted to his applying pressure on his index finger. "lucky?"

Paul took the Leatherman from his jacket, and opened the smaller blade. He wiped it with the alcohol-infused cotton ball. "Josh, this is going to hurt a little, but hang tough." He gently touched the blade to the skin.

Kendrey jerked his hand back.

"Look away, Josh."

Kendrey turned his head.

Paul once again gently pressed the blade against the side of the finger while pinching Kendrey's right inner wrist. Kendrey moved his finger slightly, then put it back. Paul quickly thrust the razor-sharp blade into the soft flesh between the knuckle and first joint, then deftly twisted the knife blade to the left of the joint. The pellet came out, bouncing sharply on the counter.

"Ow, fuck. That *hurts!* That hurts-that hurts-that-HURTS—" screamed Kendrey.

Paul ignored him as he pinned Kendrey's wrist to the counter with all his weight.

Kendrey wriggled. "Let go, you, you fascist!!"

He calmly dabbed the wound with isopropyl alcohol. "No, just an Eagle Scout."

Ignoring the squeals, which increased in pitch with every wrapping of the gauze, he finished applying the dressing and taped one side of it with one thread of tape.

"Hurts-that-hurts-that-*hurts.*"

Suddenly, above Kendrey's whines, he heard a faint 'thwack-thwack-thwack.' "Shut-up, Josh!" He put Kendrey's hand down and ran to the window.

A helicopter flew into the distance, southwards.

"Shit, trouble. Quick, Josh, follow me."

"But my, my wound!"

Paul returned to the sink, tore off a strand of tape and hastily taped the bandage onto the hand. "I'll take this stuff, and we'll re-dress our wounds really well later." He scooped up the tape, the dressing, and the alcohol and ran outside to the BMW. He opened the back door, threw in the gauze and alcohol and took the Uzi and bat out.

Kendrey followed, pressing his dressing into the fresh wound, to stop it from bleeding.

Paul glanced at his watch. Ten past four. The helicopter guys had seen them arrive, no doubt. He figured he had, conservatively, a ten-minute head-start on the thugs. Four minutes had gone by. That left six minutes.

"Let's run, Josh."

Paul set the pace, as he ran the hundred twenty yards towards the line of trees, behind which the helicopter had risen. He didn't run in a straight path, he weaved to the left, then the right; there could be a gunner waiting for them.

Between the chomp-chomp sound of his footsteps on the snow, Paul heard a plaintive whine, a high-pitched howl, to the left side of the line of trees now forty yards in front. He veered in that direction. He discerned a small dog's whiney bark.

Paul stopped next to the last tree of the line between the clearing and the next open area.

Kendrey filed in behind him.

Paul handed him the bat and peered around the tree.

A little dog, stood next to a fallen body. Another, larger dog, prostrate, attempted to lift its head, but failed. A dark line was etched in the snow next to the body,.

Finger on the trigger, Paul advanced, with Kendrey a step behind him.

"He's wounded!" yelled Kendrey.

The sheltie lifted its head, and looked at Paul, but returned to stare at his master's limp body. The dog barked soulfully, then whined. It got up and shook off the snow. It looked on silently, as Paul approached the body.

He crouched down and checked his pulse. There was none, although the body was still warm. He took his right glove from his coat pocket and put it on; he reached into the back right pocket of the trousers and took out the wallet. He removed the driver's license.

"It's him all right." He put the driver's license back into the wallet, and the wallet back into the pocket. He shook his head.

"It's so sad. The dog is so s- sad." Kendrey's jaw locked for a second before he continued, "Why do people have to kill?"

Paul turned to the other form on the white snow. He recognized a fallen buck, a sight he had seen dozens of times hunting with his father. He put the Uzi down.

The congealed blood from the buck now looked black in the dwindling light. *We can't take the body back. Should I call the police now? No. They'll ask me, and we'll now be linked to this, too.*

Paul got up and snapped photos of the body with the BMW-driver's mobile phone. He walked to the buck and did the same. He sent them to his own phone.

"Let's go. Nothing we can do here." Paul looked at his watch. Two minutes had gone by. "Josh, we gotta run. The hoods could be here anytime now."

Kendrey nodded, but he was still trying to pet the sheltie, which bared its teeth again. The sheltie looked up at Paul.

"I will . . . I'm just so depressed," Kendrey said.

Paul waited for a minute longer. He looked at his finger, which had started to throb again. He'd felt the pain of sprains and strains before, but this was different. He tried tensing his left ring finger, but it didn't move. It just stuck out now, in what might look, to the casual observer, like a perpetual obscenity.

He contemplated the next move. It led to the journalist, the last repository of information. *But if they knew to come here before us . . . they've read the article.*

Suddenly, from between the trees, Paul saw the light beams bounce upwards and then downwards, as the black car hit a bump at fifty miles an hour.

Paul crouched down and held Kendrey firmly by his right arm. "Down!" he yelled, jerking Kendrey's right shoulder down.

Within five seconds, the Sedan had turned the corner and stopped in back of the BMW. Two men emerged—one from the passenger's seat in front, and the second from the back seat, guns drawn.

Paul recognized them immediately. *Only two. Attrition. Good.* The men ran to the BMW and glanced inside. The shorter of the two—the leader—pointed to the house and put his finger to his

134

lips. They moved stealthily to the front door, guns drawn. The light emanating through the window appeared bright now.

The leader put his hand on the doorknob, turned it and leaped in, legs apart and gun held up, panning the dining room to the right. The second one followed right behind him, turning sideways and moving to the living room on the left. Paul could see him through the window.

"Where do we go now?" asked Kendrey.

"The last link is the journalist in town."

"What's her name again?"

"What does it matter, she's not famous."

His thoughts turned to this situation. The Explorer was parked about ten yards behind the Beemer, leaving enough space to squeeze it out. "Listen," he turned to face Kendrey. "Take the bat, and follow me." He took ten steps towards the line of trees and stopped, making sure Kendrey was right behind him.

"Stay low, Josh—we can't let the driver see us, but keep running."

Paul figured it was dark enough that nobody sitting in the car would notice them running from the back. And at about thirty degrees outside, no driver would keep his window open.

He ran, crouching down, from the forest, through the field, to the fence bordering the driveway. The wind picked up just as he ran thirty yards. The cold chilled his lungs. He suppressed a cough. Another thirty yards and his legs were starting to tire. He wanted to straighten up, but thought that would be fatal. He glanced back at Kendrey, now twenty yards behind him.

He was running without crouching, straight up.

Paul turned to the side and waved down.

Kendrey stooped lower.

Paul turned back to the front and out of the corner of his eye, he noticed the leader walking slowly in the living room of the house. The last twenty-two yards to the back of the Explorer felt like a ten miles. *Could he see us? If so, he'll honk. They'll come out. I only have what . . . six rounds—seven? Not more. I'd have to shoot two guys?* He realized it would be challenging unless they came out slowly, one at a time. *Slim chance of that.*

Paul executed a running slide behind the Explorer. His hip hit the ground hard. He cringed, but kept the Uzi from hitting the ground.

Kendrey followed.

As Paul caught his breath, he heard Kendrey's lungs working twice as hard as his. He waited ten seconds. Nobody had seen them.

"Listen, Josh," Paul took two deep breaths, "I have a plan."

Kendrey nodded. "Turn ourselves in?"

"Stop it!" he whispered, dead serious.

***

Through the kitchen window, Antoine surveyed the back of the house. Nothing. He walked briskly out of the kitchen and down the hall, gun drawn. He pushed open the guest room door at the end of the hall. He peered in. Nobody.

He grabbed the doorknob with his left hand, jerked the door forward and faced the space behind the door. Nothing.

Suddenly, Antoine heard the sound of a door creaking open upstairs. Parkett looked at him, pointing his index finger up. Antoine moved towards the stairway first.

He started up the stairs. Antoine stepped slowly, deliberately, to avoid extra creaking on the seasoned pinewood stairway. Parkett followed four steps behind.

***

Paul crawled until he was sitting next to the left rear passenger's door. He looked back. Kendrey was right behind him, bat in hand. He glanced at the living room window in the house. Nobody there. He rose up.

The driver was facing straight, his eyes transfixed on the barn ahead. Paul edged forward until he was adjacent to the driver, rose up and pointed the barrel of the Uzi towards the driver's head. He waited.

The driver still hadn't seen him. He took the flashlight out of his coat. He turned it on, lighting the Uzi barrel.

The driver swung his head to the left and froze. Paul gestured for the driver to lower his window. The window lowered.

"Any sound, and your brain decorates the dashboard. Now slowly, put your hands behind your head."

The driver put his hands behind his head.

Paul opened the driver's door. "Get out. Drop down your hands and knees."

The driver got out of the car and awkwardly knelt down, keeping his eyes fixed on the barrel of the gun.

"Josh, your turn," Paul called, quietly.

Kendrey nodded, and whacked the driver in the back of the head with the bat.

He fell to the ground, out cold.

"Good job. Now, Josh, open the driver's door of the Beemer. Then get in on the other side. Quietly. Go."

Kendrey stood up and ran, crouching the thirty feet to the BMW's driver's door.

Paul went around to the other side of this SUV, the side facing away from the house. He once again pulled the Leatherman from his inner coat pocket and pulled out the razor sharp short blade. He perforated the right front tire. Gases, smelling of rubber extracts, filled the air. He did the same to the right rear tire. The Explorer gently sank to one side as the air hissed out of the tires.

***

Antoine stood tall at the landing at the top of the stairs, listening for the slightest sound. None. He crouched down, undid his laces, and slowly, deliberately removed his shoes. Parkett, now also at the landing, did likewise. Antoine stealthily approached the open door of the first bedroom, gun up. He stepped inside, targeting the closet. Suddenly, he heard what sounded like the toilet seat wobbling on top of the ceramic bowl, somewhere in the next room. He turned around, bringing his gun down and signaled to Parkett, who was still in the hall.

Parkett stepped next to the bathroom. The door was six inches ajar. Antoine took his position just to Parkett's left.

137

Parkett kicked the door open with his right foot, simultaneously swinging his gun in.

A cat, standing on the toilet seat, jerked its head up from inside the toilet, smacked her lips and arched her back. She hissed and bolted out between Parkett's legs.

Antoine heard the BMW engine rev up.

Within five seconds, he had slipped on his shoes and was sprinting down the stairs. He left Parkett fumbling with his shoestrings. Antoine ran out the front door. The BMW was forty yards down the driveway. He knelt, firing the pistol three times. The glass shattered as a round pierced the back window. He may have hit the driver, but he wasn't sure.

***

Paul looked at the fresh hole in front windshield, then straight ahead.

Kendrey lowered his window. "I'll get those assholes," he screamed. He reached back and took the Uzi with his left hand.

"Wait, Josh!"

Kendrey ignored him, as he swung the weapon out of the open car window, holding the muzzle with his right hand. He squeezed the trigger with his left hand, but didn't hold the handle firmly. Six rounds went out, as the Uzi kicked up, bouncing Kendrey's forearms against the top of BMW window frame. As the BMW turned right around the bend, the Uzi fell silently on the snowy driveway behind it.

"Hey, what are you doing?!" yelled Paul.

"Oops!" yelled Kendrey. "Want to stop the car so I can pick it up?"

Another shot hit the right, then left rear passenger window.

Both Paul and Kendrey ducked.

Paul emerged, keeping the car going forward. "Shit, Kendrey, could you be goddam normal for fifteen minutes?!" Paul yelled as he throttled the gas.

A minute of silence followed like a marble slab headstone in the dead of winter. "I'll t-try," said Kendrey sheepishly, "real hard."

138

***

Antoine turned Luke's face up. He slapped him across the cheek. "Luke!" Antoine yelled. No response.

"He's out. Parkett, open the back door," Antoine ordered, as he lifted Luke up and moved him towards the back seat. "You drive."

Parkett started the car and backed out. The Explorer veered to the right, straight into the doghouse next to the driveway.

Antoine heard a loud crunching sound and the car stopped. "Parkett!"

"Boss, it's not me. The car's . . . feels like a flat."

Antoine leapt out of the car and looked at the right tires. He started kicking the front one. "Dat lousy fuck!" Antoine started kicking the snow in all directions. He looked over the fence, then all around the perimeter of the property. He couldn't see any lights. He ran to the right twenty yards.

Then he remembered the note he had seen. Stoltzfuss wasn't alone – his wife would be back soon. He walked back slowly.

"John, It's okay. In fact, I got an idea ta turn 'dis shit into manure. We just need ta find the body. Here's the plan . . ."

21

Paul glanced at Kendrey's hand, which still wasn't properly bandaged. The gauze was sticking only due to the coagulant in the blood that coursed through the shot-up veins. He started feeling remorse for having asked Kendrey to appear "normal." *Is he less "normal" than anyone else?* "Josh, I'm sorry about that comment about you being normal. It was dumb of me."

"That's okay, Paul. People say th-that all the time," Kendrey said with a faint smile. His jaw locked. "B-But I'd like to call my mom."

Paul nodded. "The lawyer's phone was full. *Maybe he hasn't blocked it yet.*"

"So where are we going now?" Kendrey asked.

"Hollidayton. About ten miles up this road. We're gonna pay the journalist a visit."

"No signal."

Paul shrugged his shoulders.

Four miles later, Kendrey dialed again. "Hi, Mom. Yes, I'm still out of town. But the reporter said we'd finish up soon. We're just going to visit the journalist who wrote the story about the guy who thought the machine ch-changed his vote," Kendrey said, now a little tired. "I'm fine . . . but we saw a dead buck. It was sad . . . Okay, so you'll be up? . . . Then I'll call later. Bye."

***

Latanya punched the bag one more time. She hadn't trained in over eight months and it felt wonderful. She launched a volley of punches, dropped backwards and delivered an explosive roundhouse kick to the Everlast bag suspended from the ceiling. It moved a foot then jiggled back and forth for a few seconds, before resuming its circular twist.

Mixed race marriages were renowned for being more strenuous. It wasn't surprising. Different cultures. Circles of people that had different degrees of cross-racial acceptance. . . in both directions. But she didn't want to bow to her Mother's pressure. *Bottom line. Do I believe in him enough to give him another chance?*

"Not bad, not bad. Now let's see some drive behind it," her trainer barked.

In a smooth arc, Latanya hit the bag high. The bottom moved less than last time. Gotta do better than that.

"Come on, come on."

Latanya nodded. She punched the bag with a flurry, crouched down and with an explosive jump, twisted her entire body 120 degrees, her right leg trailing. A perfect hit. The bag lurched up, then two feet sideways. She thought she saw a puff of dust on impact. YES! *Okay, I'll give him one more chance.*

"That a girl!"

But I'll have to do this tough love hold-out for a few days longer. He's got to think it's over.

***

Antoine cupped his hand to his ear. "She's here." He and Parkett crouched next to the fence, near the start of the driveway. As Stoltzfuss' wife passed the gate, she slowed down to turn left onto the long driveway.

Antoine rose up, stepped into the middle of the driveway and started waving his arms, signaling for her to stop.

Mrs. Stoltzfuss stopped the car, but instinctively pushed the lock button down.

Antoine approached the driver's side window. He pulled out the first of his two badges, the State Police one, and put it to the glass.

Mrs. Stoltzfuss lowered the window an inch. Antoine placed a gloved hand on the lowering glass. "Mrs. Stoltzfuss?"

"Yes."

"Officer Batulli, Ma'am. An accident took place here a few hours ago. We believe it might'a involved ya husband."

Mrs. Stoltzfuss placed her hand over her mouth, in shock.

"We'd like ya' ta take a look."

Mrs. Stoltzfuss approached to within ten feet of the body then stopped. Tears welled up in her eyes.

"It's him?"

Mrs. Stoltzfuss nodded. She put her head down and sobbed.

Parkett stepped behind her. He looked at Antoine.

Antoine nodded.

Parkett placed one hand in the middle of her back. He stepped to the side and placed his gloved palm on Mrs. Stoltzfuss's solar plexus. He pressed both hands together, squeezing hard. The veins in his head swelled as he pressed.

Three seconds later, Mrs. Stoltzfuss collapsed.

"Good," Antoine said." Now John, let's cover our tracks," he continued, as he took a step backward. He bent down and brushed snow on his imprints. He stepped back again and repeated. Parkett crouched next to him, doing the same.

Antoine first opened the trunk of the Explorer. With Parkett, they transferred Tom's corpse to the trunk of the Pontiac.

Luke was now standing.

"You look wobbly, but listen, Luke. You'll bring the the car back after we get to town. But ya gotta wipe down the inside. Destroy all fingerprints. Now get in the back."

Antoine opened the passenger side door of the Pontiac and helped Luke in.

Parkett started the car, then stared blankly ahead.

"What are ya' waitin' for, a funeral procession?" Antoine barked. *"Let's go!"*

***

Paul drove past the "Duncansville City Limits" sign, and he relaxed for a second. But as the car passed a couple of people walking on the road, they stared at the BMW. Paul looked in the rear-view mirror. They were now pointing. They passed a bus stop. "I don't like this—we're going to be ID'd. I know these towns, and it's a big thing to see a Beemer here, let alone a black one with bullet holes. We're going to have to abandon this thing before we get into town."

As he slowed down and turned the BMW around, Paul turned to Kendrey. "Wait Josh, if they got to Stoltzfuss, they'll know

142

about the journalist! What's stopping them from sending another car?"

"Concern for climate change?"

Paul couldn't suppress his laughter. When it was finished, a half-minute later, he turned to Kendrey. "You say the damn-dest things." He then suppressed the smile, pursed his lips and gunned the accelerator. They passed a bus stop. He picked up his phone and pushed a speed-dial number. He had driven over a hill. In the distance, the center of a small town was visible. *McKee.*

"Come on, Mazurski, don't be a pussy."

On the fifth ring, he picked it up. *"Buddy, can't talk. Sick as a dog."*

Paul heard the sound of vomiting, and the call was dropped. "Shit," he exclaimed.

Paul then slowed down as he typed an SMS. "Warn jrnlst O'Reilly, Altoona Bnd, life in dgr. Go 2 police, wait4me." Paul pushed on 'send.' The iPhone screen went black, with a circle of bars in the middle. *Shit.* Paul glanced at the road, then picked up the BMW driver's phone. He checked that it was on, and pulled over. He composed an SMS, pushed a number, waited until it connected, and sent it.

"It's good we're going slow, I'm getting carsick."

***

Antoine was fixated on the road ahead and the next set of red lights as their Pontiac drove past a bus. A large sleek black car appeared under a light ahead before disappearing in the dark.

"Speed up, Parkett, target may be at three hundred yards." Antoine brought his Glock level with the glove compartment. As they approached, he put on his reflective sunglasses. Seventy yards...fifty...thirty.

"Let up, Let up. It's not them. It's a Lexus."

***

Paul saw two families walking along the road. They looked like Mennonites. A bus stop appeared in view.

143

By now, Paul reasoned, whichever organization the thugs worked for would have tracked down the phone and he had made as many calls from it as possible. It didn't help that all the people he tried to call were protective of their privacy and most wouldn't take calls from outside random callers. He had one more call to make.

He dialed the number he got off Google.

"Hello, Hollidayton Police Department."

"Hi, this is Paul Robinson, reporter from WQHP. I wanted to report that a journalist, O'Reilly, is in grave danger."

Kendrey suddenly grabbed the phone.

"He's not K-k-kidding. Grave danger. There's gonna be a ton of graves needed for all the people killed today. Yes s- s- sir!"

"JOSH! What the fu--!" Paul yelled, as he grabbed the phone and slammed on the brakes.

"Wait, who did you say you were?" the operator said.

"Paul Robinson, reporter from WQHP."

"Just a second, let me get the Captain on this."

Paul glared at Kendrey. *Un-f---ing-canny.* A full half-minute passed and Paul started wondering why the delay on the other end.

"Yes, Mr. Robinson," a man on the other side finally answered in a tenor voice, a short person, no doubt.

"Yes, I wanted to report that the journalist O'Reilly—"

"The cute one!" yelled Kendrey.

Paul quickly covered the phone with his left hand, turned around and punched Kendrey in the shoulder. "Shut-up!"

"Ow!!" Kendrey looked at Paul, his eyes wide open.

"Kendrey – Normal!"

Kendrey pouted, staring back at Paul.

Paul lifted the phone to his mouth again. "Yes, is in mortal danger."

"Uh-hunh, Mr. Robinson, and can I enquire where you are calling from?"

"Yes, I'm on Route One-sixty-fo—" *Why would he ask me where I'm calling from? Oh shit, of course, Because I'm a fugitive! And now they know where.* "Uhm excuse me, yeah Dunnings Highway, from

Claysburg." Paul lied. *Tracing this phone!* "Power on my phone's going, sorry, offi—" Paul shut the phone.

"The power wasn't going off on the phone, it's almost a fu-full charge," Kendrey said, coolly.

"Just clam up, Kendrey!"

"Silly expression."

Paul wiped down the HTC quickly against his trouser legs. He slowed down to about ten miles-an-hour, rolled down his window and threw it out. "Too bad – it is a neat phone."

Kendrey started gorging himself on chips. "Mmm. Nyami, nyami," Kendrey mumbled, as the chip crumbs spilled out of his mouth. "Oh so *good!*"

Kendrey turned to Paul and said: "Paul, this picnic's been lots of fun and all, but I wanna go home."

"Josh, these guys will kill half the town to keep us from leaving alive."

22

Antoine held the mobile phone to his ear. "How close are they? We should have caught up by now."

"Judging by the last signal, about three miles outside of town. Adams Road, specifically."

"I didn't see it. When are the others coming?"

"They could be there already, fifteen minutes away, tops."

"Fine, I'll meet up with 'em."

Antoine peered again down the road. He saw more and more houses and lamps lighting driveways. But he was focusing on the cars.

***

He usually managed to use the phone two more times after it had gone dead the first time. Paul speed-dialed Latanya's number. No answer. When the voice messaging prompt came on, he didn't hesitate. "Tan, the voting machines. Someone is killing—" Paul's cellular made the 'bling-bling' sound and went dead. Paul hit the phone against the steering wheel.

"Goddamn battery!!" He yelled.

"Fifth generation will run for a week on a charge."

Paul calmed himself down with the thought that they just needed an outlet for his phone and access to a computer.

"So Josh, where would the bad guys be waiting for us?"

Kendrey started squirming. "Listen, I've got lice on me—" Kendrey sputtered, before his jaw locked. "And . . . and they're going into my open wound. Aarghh!"

"Josh. Calm down. Take the medication."

"I need to call my Mom."

"Josh, we'll have to get to a hotel."

"Mom's going to worry about me!" Kendrey yelled. "The vermin of the earth are now on my arm. Arghhh. They're in my veins."

"Okay, okay."

Paul swerved the car into a forest road to the right. He drove a hundred yards in and killed the lights.

"How are we getting into –"

"By bus. We just passed the stop."

***

Dr. Artois was riding in the taxicab, on the way to the studio, when she heard the ping of another received message. She picked up her Blackberry. The screen read: 'Stoltzfuss Killed. Election machines.' *Code?* The number was not familiar. Nor did she make any association between this statement and anything she had committed to doing this week. Nonetheless, she dialed the number.

"*You have reached the answering service of Attorney Crossman—*" Wrong number. Her message box was full, so she deleted the mis-addressed SMS. She looked at her watch.

The taxi had just hit a small traffic jam.

She wondered what flora had caused her colleagues at WQHP to be so sick. Her presentation wasn't going on until 8:00 and she was going to go in at 6:30 anyway, but until now she had made sure that the wall between her work at the Harrisburg General Hospital and the station was very solid. Unlike the omnipresent CMN medical consultants, she made almost her entire income from the real job. She had a computer and a cubicle at the station, which was all she really needed. And the service she presumed she was about to provide was also going to be pro bono, or paid—woefully—through declaring a few false overtime hours. But she didn't mind, she'd always considered providing medical advice and pointers a service to the community, the same community that had once bought Thin Mints and Samoa's from her. She looked once more in the satchel she had filled and taken in addition to her handbag. Activated Charcoal, Pepto-Bismol, Smelling Salts, hydrocortisone in the event of shock.

Of the spate of possibilities, the only one that concerned her was e-coli. That could have been from salad, but the onset was usually much slower. She figured that most of the morning crew had lunch at around 10:00 or 11:00 in the morning. This meant

147

that onset wouldn't be until this evening. She scratched her head.
No, this was most probably salmonella. Breakfast at 10:00 or
10:30 would coincide with onset of symptoms about now. But
that was strange, she thought, since the five egg producers to
which the spring outbreak had been linked were surely inspecting
diligently now. So would the USDA. *Another egg producer or tomatoes,
apparently.*

***

As he approached the main intersection to a highway a quarter
mile up the road, he slowed down. Paul looked out the window.
He saw another police car following the two that had passed him a
few minutes earlier. Since the Stoltzfuss farm was some ten miles
east of his current location, he had some time. The police would
find the bodies and that would buy time.

He spotted a hotel, "The Americana," a simple three-story
fleabag. *Perfect.*

They got off at the next stop. The front of the hotel boasted a
worn Penn State flag and two faded beer ads. One of the two
outside lamps was burned out.

"At least they'll have a b-bathroom," Josh said.

"Yeah."

Paul and Kendrey entered the hotel and walked ten yards
down the felt-papered hallway, to the counter. The countertop
was polished, while nothing else appeared to be; dust balls of
varying sizes rested in every mailbox. A man appeared, his face
pock-marked and his long greasy hair tied back.

"Now if I could see a major credit card," he said.

"We'll be paying in cash."

"It's policy, you know, in the event of damage done to the
interior . . . "

"Of this wreck? Fine. Could you just run that through an old-
fashioned paper imprinter?"

The clerk raised his right eyebrow, "Whatever," he replied,
shrugging his shoulders.

They were given the "only available" room, on the top floor.

Paul was struck by how pockmarked the 1980s furniture looked with cigarette burns. One nightstand had a deep scratch on the side, cutting through different and sundry marks. The lighting gave him the impression that a séance was about to take place.

"Go ahead and wash up, Josh, I'll charge my phone, make some calls."

Kendrey stormed into the bathroom and shut the door. He turned on the faucet.

"A-a-ahh, finally," he heard from the bathroom.

He took the charger from the backpack and plugged it in, connecting it to his mobile. He then picked up the hotel phone receiver and dialed '9-1-1.'

"Emergency. How may I help you?"

"I'd like to report a murder."

"Please identify yourself."

"This is Paul Robinson from WQHP," he said, as he crushed an adolescent roach with his left shoe.

"Excuse me, uh, Mr. Robinson—wait are you the one that did the story about the golfing chimpanzees?"

"The same. Listen, it's . . . "

"Yeah, I loved that. So are you reportin' a murder, or are you asking us if we have any murders to report? Or is this another prank ca—"

Paul jerked the receiver away, as a high-pitched tone screamed through the phone. He held the telephone a foot from his ear. *What the hell was that?*

Paul dialed Watkins' number. He waited.

The lyrics "A-ma-zing Grace, how sweet thy sound," came wafting through the crack under the bathroom door.

Paul got up and ran to the bathroom. He gestured, sliding a finger across his throat and part-closed the door.

"Pity a poor soul like me . . . " Kendrey continued, ignoring the request.

"Goddammit!" Paul scratched his head. He glanced at the photocopied reproduction of Van Gogh's 'Café at Arles' as he walked to the window and pulled the curtains apart.

He lunged from the window to the bathroom. He kicked the door open.

Kendrey sat in the bathtub, surrounded by thick bubbles. The foam on top of his head was in the shape of a boat. "Admiral Nelson. You like it?" Kendrey asked.

"Get out *now*. They've found us!"

"But I'm, I'm just starting to get clean."

Paul hit off the bubble hat, grabbed Kendrey by the hair and pulled his head up to chest level.

"Ow! Okay, okay!"

"In ten seconds, we are out that window."

Paul bounded out of the bathroom. He crouched down, disconnecting the charger from the plug. "Those guys, they're all over the goddamn place, like that Agent Smith in the Matrix."

Kendrey emerged from the bathroom, toweling off madly.

"But now I'll catch, I'll catch my death of cold."

"Josh, if you want preventive lead therapy, stick around another minute."

Paul opened the window, looked out and saw the sidewalk thirty some odd feet below. He stuck his head back in and looked around the hotel room. Kendrey was putting on his pants.

***

Antoine and Parkett rushed into the hotel lobby. No receptionist. Antoine hit the silver bell. The man appeared ten seconds later. Antoine saw a tattoo on his upper wrist, long hairs coming out of the guy's nose and a certain gait – not confident but not fawning either. The guy had been in the slammer. No sense in pretending to be FBI or Police—this clerk would see through it.

"How can I help you guys?" the receptionist asked, brusquely.

"We're here to meet wit' a friend, a Paul Robinson."

"Hmm. He said he didn't want to be disturbed. And we have a privacy policy—"

Antoine whipped out his gun and pointed it at the receptionist's forehead. "Your privacy policy again?"

"Don't shoot, be cool man. Room three-two-three."

Antoine lowered his gun. As he turned to go up the staircase, he saw the receptionist eyeing the phone.

"Now *my* privacy policy—if ya' call the cops, I remove your privates. Clear?"

The receptionist nodded.

"Good. Oh, and in case ya' wondering, all the lines to the cops are tapped, like you'd ever be friends wit' da Law."

The receptionist kept his hands up and nodded in resignation.

With his Beretta drawn, Antoine climbed the stairs, followed by Parkett. They stood on either side of the door. Antoine put his hand on the door knob and gently attempted to turn it. Locked.

Antoine stepped in front of the door, gun out and kicked it open. He saw a fast, fluttering movement. He pushed lightly on his trigger, then let up. The curtains fluttered again as the wind sailed through the open window. The bed, the corner of which was lifted about two feet high, was jammed against the window frame.

Antoine ran to the window. A sheet was tied around the leg of the bed. He looked down the sheet to see another sheet tied to the first one. *They couldn't have taken a car, they must be on a side street.* "Let's go," Antoine said as he turned around and ran down the stairs.

"Fine hotel. I'll come back for ma second honeymoon," Antoine yelled as he passed the receptionist.

23

Paul and Kendrey ran up Blair Street. Paul saw the street curving to the right, gradually increasing in grade. "Hug the left side," he yelled to Kendrey. As they approached the main intersection, he saw a coffee shop. Another step, and he spotted a "New Wi-Fi Hotspot" decal in the window. He stopped in his tracks. "Yes-s-s, we're good," he muttered as he glimpsed the open notebook deep in the back of the I. "Here!" he blurted out, as Kendrey ran past.

The two burst in. Kendrey said, "I'm freezing. Can I have a cappuccino?"

"You're nuts, really."

"Ple-e-e-ase?"

"Okay, okay," Paul relented.

He continued to walk towards the young teenager with the laptop computer, sitting in a nook. He glanced at the screen "Cool, you've got Quake Four."

"Yeah. This guy is giving me a post-Viliagra let down. Can't play this game with neophytes."

The kid hasn't even reached puberty, what is he doing talking about Viliagra. "Listen, uh–"

"John."

"John," Paul flashed a super-smile. "I've got an offer you can't refuse. If you let me use your notebook for five minutes of e-mailing, I'll give you ten bucks."

"Dude, I'm here to play online. If I needed money, I'd be at home washing dishes," the teenager replied, without looking up.

"Fifteen bucks."

"Twenty. Money up front," he retorted without looking up.

"Deal," Paul said, before looking through the window. Nobody stirred. *They'll only find us if they go street-by-street. And they won't, if there are only two . . .*

The teenager set up the Firefox portal. Paul took out two bills and handed them over. He sat down in the nook, and started typing. 'Maggie, the Danabold voting machines were rigged. A big conspiracy. Witnesses being killed. Stoltzfuss, Millersburgh, Blair County, killed.'

Kendrey put Paul's cup down. "Thanks," Paul mumbled as he typed "M. Leboeuf, Duprix [third guy] witnesses" Paul heard the door closing. *No way. Too soon, okay at least two…* He clicked on send. "John, what's your e-mail." John spelled it out.

***

Antoine had almost reached the fire station when he received the radio message. "Found them. Panda Coffee. Twenty Elm. Place is full."

"Roger. Just bring 'em out – use your police badge." *He should be okay.* Antoine's cell phone rang. He took it out.

"*We're in town*," the caller said.

***

Kendrey turned around suddenly.

"Where is the reporter?" Paul heard the thug ask.

From where he sat, Paul could only hear him. Kendrey obstructed his view entirely. But this worked both ways. Paul glanced at the steam coming from the coffee.

John involuntarily glanced at Paul.

That's it… won't be precise, but an ounce in the right spot will do.

"Oh, you mean the spaceman? Ground control to Mister Paul," Kendrey sang out.

In one motion Paul threw the coffee onto and over Kendrey's right shoulder.

"Aargh!," Paul heard. He jumped up, squeezed between the table and the wall.

The thug was bent over, covering his face with his hands.

Paul lifted up a chair, and brought the edge of the seat down on the back of the man's head. The man collapsed in the long puddle of coffee.

Paul waited three seconds. The man was out cold.

Kendrey stood still, holding his Latte.

153

Paul grabbed his backpack. "Come on, Josh, we gotta run. He grabbed Kendrey's shoulders and shook him. He turned to the teenager. "John, send my e-mail to all your friends, too, bud. See you around."

The teenager, wide-eyed, nodded slowly.

"Oh, and kid, we're the good guys," he exclaimed as he and Kendrey passed the end of the counter.

The staff at the café stood like statues, as if they had been playing freeze tag.

Kendrey gestured for a lid.

The barista finally moved. In reaching for the lid she knocked over one of the two piles, then offered a cap from the intact pile to Kendrey, not taking her eyes off Paul.

"We need to go out the back."

The cashier opened the swinging gate and gestured behind him.

"Whoa, Whoa!" the shift manager exclaimed as Paul and Kendrey barged into the office.

"Just passing through . . . I'll need an envelope and paper, too. Thanks," Paul said and snatched a large package of eight-and-a-half-by-eleven-inch white envelopes from a pile to the right of the desk. "Oh and call the police. Tell them journalist O'Reilly is under threat—"

"Yeah, the really hot one," Kendrey interjected.

"Shut up, Josh!"

The manager scratched his head as they headed past him.

Paul had no time.

They exited ran to the left, then turned right at the end of the alley. It was a North/South alleyway. Paul knew they were going the wrong way, but for now they had to get as far as possible from the Cafe. And they had to drop off the information to the police.

**

Antoine burst into the cafe. All twelve customers looked up at him. He produced a badge and held it up high moving it so all could see.

"State Police. Nobody move. We're pursuin' two escaped terrorists, one of which is also a confirmed rapist. Has anyone had any contact with them? We're talkin' brushed past you, used your cell phone . . ." He scanned the room. He saw some notebooks out. "Or laptop," he concluded.

An elderly lady turned her head, exaggeratedly, to the young teen at the back of the shop.

"Thank you, ma'am, your patriotism will not go unnoticed."

Parkett was out cold. Antoine stepped over him and stood, towering, over the teenager.

"Hello, I'm Inspector Johnston. Could ya' tell me what happened here?"

"Yeah, I was playing Quake, and these guys came in. One of 'em offered me twenty bucks if I'd let him play, too."

"What's ya' name, young man?"

"J-John," he replied, nervously.

"Ah, just like the apostle. John, could you please come with me. Take ya' laptop, we'll need to dust it for prints."

The teenager followed behind Antoine. Before exiting, Antoine turned to the shift manager, now standing next to the cashier. "Which way did they go?"

The shift manager pointed behind him and replied, "Through my office, out the back."

"Which way?"

The shift manager shrugged his shoulders. "D-don't know."

Antoine stepped out the front door.

A black Explorer pulled up to the curb, past the door of the Panda Café and three guys piled out.

As Antoine turned towards them, he heard fast footsteps. He turned and watched the teenager sprint. He was thirty yards away when Antoine lifted his gun and took aim. He lowered his arm and turned towards the Café. Eight pairs of eyes were affixed on him. Antoine smiled and took a deep bow.

People in the café smiled. Some chuckled. One clapped.

Antoine turned towards the three agents. *Who's the fastest?* "Leroy, find that kid and take his laptop." The "inspector" with the physique of a pro football player took off.

155

Antoine faced the second one. "Dan, go to the police station and stake it out. Don't let that reporta' go in. Luke'll give ya' directions. He's in the Pontiac," he said, gesturing with his hand towards the car across the street. *Who's the second fastest?* He turned to a thinner, but still muscular agent. "Jimmy, they just took off out the back. You patrol in that direction—stick to the main street after 'dat. We don't want the sonofabitch phonin' from someplace." Antoine pointed north of the café. "In twenty minutes, no more, meet at the journalist's house."

Antoine faced the driver. "Tony, ya got the two spares for the Explorer?" he asked.

"Sure-they're in back."

"Good. Now pick Parkett off the floor and put him in the car. I don't know where that reporta' gets his luck from, but I was wrong. He doesn't have the luck of an Irishman . . . he'd make a fuckin' leprechaun jealous."

Antoine then stepped off the curb, into the street.

'Special Assistant' Frank Rayman sat in the back of the car, watching everything with a detached air.

Antoine could feel his blood pressure rising. He turned towards the Pontiac and signaled for Luke to get out and join him in the new Explorer.

Antoine climbed into the driver's seat, and Luke joined him in the back seat. Antoine turned to Luke and started explained what he was to do next. "Call me when you guys are ten minutes from here. We'll meet ya on the Route in back. Got it, Luke?" he concluded.

Luke nodded. The door opened and shut.

Antoine turned to the man crouched in the seat. "Frankie Rayman, the two of us, we're goin' to a hot babe's house. I got her all steamed n' ready for ya."

"But I stopped—"

"Shuddup, Frank. We know about the thirteen-year-old in York. What? She asked ya' to mount her in exchange for buyin' Girl Scout cookies?"

Rayman was silent.

"Look, Frank. You know, that's my name, too, right? What a coincidence." Antoine smiled broadly.

156

Rayman forced a fake two-second smile.

"We all got our tastes and preferences. Yours are on the young side, that's all. It's okay. Now, like Luke told ya, we need you to occupy the girl, while we check her house out. We don't want her even thinkin' about what we're lookin' for. So keep her satisfied."

"I don't know if I can. The pressure—"

"Sure you can, Frank." Antoine pulled out a small blue-green and a larger yellow pill and held them out in his palm.

"What are those?" Rayman asked, smiling nervously, as he pointed to the yellow pills.

"Well ya' know what da little blue-green pill is. And the yellow one next to it, is his longer-lastin' cousin."

Rayman nodded. "Thanks, but I don't really . . . "

"Just take 'em. Don't dishonor me," Antoine said, smiling, and clasping Rayman's shoulder firmly in his oversized hand.

Rayman cringed. "Okay . . . have any water?"

"We'll get it on the way. Anythin' else? Twinkies maybe?"

***

Paul and Kendrey were huddled under a tarp, next to garbage cans. Paul heard the footsteps just on the other side of the fence. Through one of the eyelets, between the fence cracks, he saw the thug's feet. They turned to the left, then the right. They turned towards them. The thug got on his tiptoes.

No . . . no.

The thug turned his feet to the left and ran.

Shit, I forgot to put that period between the names in the kid's e-mail address. Paul turned to Kendrey. "I have no clue if my contacts got the message, Josh . . . We gotta get this recording to the police."

As soon as he could no longer hear the sounds of footsteps, Paul threw the tarp off and got up. He poked his head over the fence, scouring the street for other thugs.

Kendrey was crouched down on the ground. He picked up a piece of pipe from a pile of half rusted metal.

"Good idea."

157

24

The Deputy Producer, a Mr. "Watkins" fumbled with the phone, before putting the receiver to his ear. "Hold, hold a sec, I have a visitor." He put down the phone. He smiled like it counted. "And what can I do for you?!" Watkins asked.

"I think the question is . . . what can I do for you?" she answered, with a smile. "I'm the temp you ordered."

Watkins scratched his head.

She liked this part, when absent-minded men—all the straight ones—could never remember "ordering" her services.

"Well, I hadn't ordered a temp, but it's a good thing someone did; this food poisoning is a new low for a cafeteria known for low points."

"Amanda's the name," she said smiling. She looked to the left, then to the right.

"Oh, right, Amanda. Yes, well, make, make yourself comfortable here," he said, pointing to a chair by his desk. "I'll be right with you."

Amanda put down her bag and scanned the desk. She draped her coat over the chair and glanced at the fax machine. She sat down.

"Okay, air that after the bird flu story," she overheard Watkins saying.

She looked at the desks. She had to pick up on one of the clues. This Mazurski guy bowled, fished and professed a slavish devotion to the Nittany Lions. That described what, 99% of the male population around here?

The producer was busy talking about something, so she got up and decided to meander from cubicle to cubicle, taking in the Penn State logos. Between glances back at Watkins, she vetted the cubicles, one after the other. Every time she glanced back, she could almost hear Watkins' lips smacking together salaciously. *Short skirts are a godsend in this business.*

After a few minutes, she saw a photo with Mazurski and Robinson, at a bowling alley. *That's it.* She glanced in the direction of Watkins' desk.

Discretion had apparently gotten the better of him and he was no longer looking in the direction of her upper thighs, but was still talking heatedly on the phone.

She sat down at Mazurski's computer and started typing in the passwords she'd been given. The third one worked. The network mail came on. The list of e-mails appeared. She scrolled down. 'Weather update', 'Meeting Agenda, 'Emergency-Murder'. The temp opened the last one. Maggie, Dr. Artois, Mazurski. She took a pencil and a post-it note pad on the desk, on which she wrote down the recipients. She then deleted the message, before clicking on the 'Trash' file, and deleting all. She took the top post-it note and repositioned the pad exactly as it had been before.

"Fine. Now, *that* story is dead and gone," she overheard. A phone hung up. She got up and walked, in a slow catwalk, back towards Watkins.

The producer approached in kind and she smiled seductively.

"So, uh, yeah. We have . . . files to sort," Watkins stammered.

"And I sort e-mails . . . and I'm good with hard drives, too," she said, pulling her shoulders back just a little.

"That's, that's great, I have a great big . . . problem with one. But um, later. I usually have tea about now. Let me, let me show you where the kitchenette is, Miss Amanda?" Watkins stammered.

"Just "Amanda" is fine."

***

Paul and Kendrey reached the driveway next to the hardware store. Paul looked around the corner, and saw a man in a dark coat across the street, reading a newspaper under the street lamp. Steam rose out of his mouth with every breath. The man started looking up.

Paul ducked back behind the wall. Out of the corner of his eye, Paul saw the teenager John running past this street along another alley.

"Shit. Josh. I, I got that kid in trouble!"

159

"He's a gamer, he'll sort it out."

Paul shook his head. "Josh, try to get to the back door of the station. Cross the street a couple of blocks down and wait until he's reading. Here's the envelope with the recorder and the note." He handed Kendrey the envelope.

Kendrey looked wide-eyed at Paul.

"Look Josh, I know you got this condition. But we need you. Bi-polar disorder—you know Lincoln, Kennedy both had it. And they saved this country. Now . . . it's your turn. Go!"

He looked on as Kendrey hustled to the back of hardware store, and turned left. *Giving a task like that to a guy who's certifiably nuts. What am I coming to?*

Paul heard some footsteps in the distance, to his left. He crouched down and saw an enormous man running through the same alley the teenager had just run through, about forty yards away. Paul ran along the street to the edge of that alley. He saw the agent stop and turn right. Paul sprinted down his own street. He ran across a lawn at an angle, then next to the building. He stopped, listening to the footsteps.

Click, click. *Now.* He jumped on the thug. An imperfect hit, but still managed to get the guy off-balance, and the two came crashing down.

Paul got up first, and hit the thug in the jaw with a strong punch.

The man, unfazed, got up. He parried Paul's second punch and delivered a counterblow to Paul's midsection.

Paul doubled over.

"Inspector Fleming. State Police H-Q," the man said.

Paul looked at the badge, nodded, bent down as if he was catching his breath, and then rammed into the thug head-first, grabbing and lifting the man's legs off the ground.

Fleming fell down again. It took him two seconds to kick Paul off. Paul suddenly felt himself being grabbed, lifted and thrown on his back. *Ow. Shit he's strong.*

The man was now kneeling on him, with one hand on his throat. The man pulled out a 12-inch bowie knife from a sheath attached to his leg. He positioned the knife to Paul's right ear, laying most of the icy blade next to his pulsing jugular.

The thug smiled, as if experiencing an 'Aha' moment. "Chimp reporter, you're gonna be the story now."

"Really," Paul replied, breaking into a cold sweat. He tensed his thigh muscles, trying to see if he could throw the guy. Zero chance. The guy was a boulder. Paul felt the blade pressing into his skin. "Ooh, ooh!" he hooted, at the top of his lungs.

Out of the corner of his eye, Paul saw the teenager leaping behind the agent.

"Last wishes?" the thug said.

"Redemption."

The man looked puzzled, just before a brick landed squarely on the back of his head with a 'Thwack.'

The thug fell, unconscious, on his side. The teenager beamed, as he cocked his hand and yelled, "YES-S."

"Thank *you*, John! High-five" Paul got up and they patted palms.

Paul briefly brushed the snow off his coat. "Good to see they didn't catch you."

"I ran. You were right. Dja see, the guys didn't have standard issue guns. Scary."

"That's no accident," Paul started, as he bent down and opened the thug's jacket. Do you have a place to hide?" He slid a Smith & Wesson out of the man's holster. He lifted it up and cocked it, then slid it into his waistband.

"Yeah."

"Good. Go there until this blows over."

"No way, this is cool. Haven't seen this kinda stuff . . . ever."

Paul looked both ways, then grabbed the boy by his jacket, lifted and pressed him against the wall. "Look, John, This ain't no hip-hop. These guys are like the baddest-ass homeys from Grand Theft Auto." Paul let go of the jacket and John slid down. Paul held up his left glove, the blood now congealed to almost the same color as the glove.

"See this?"

Paul waited for John to focus on the bloodied part, and the finger sticking in the air. "They shot me through the car window. A shotgun, two automatic pistols. They've murdered two people in the last four hours."

The kid's eyes opened wider. He slid down the wall a few more inches.

"So get the hell outta here. In an hour, when this clears up, go to the police. Not now, they've staked the station out. For now, hide somewhere good."

Paul glanced down the street. He thought he saw another black-coated man running across. "Have you fired a pistol before?"

John nodded. "In the Scouts, two weeks ago."

Paul took the Smith & Wesson back out from his waistband. He put the safety on. "Here John. Take it. If—but only *if* – they find you in your hiding place, click the safety off—" he demonstrated, "and blow the thugs away." Paul flipped the gun backwards and held the barrel, so John could grasp the handle.

John nodded, taking the pistol. He stuck it in his waistband and disappeared into the night.

Paul ran into an alley heading in the direction of the journalist's house, now a hundred yards farther away than it had been five minutes earlier.

***

As they approached O'Reilly's house, Antoine felt for the "tools" in his pockets. He removed the dark blue FBI cap, with the rich yellow-gold knitted letters prominent on the front. The driver drove around the back of the house. As per the plan he'd be at the end of that street, in the darkness under the trees, ready to drive off in a split second.

Antoine turned to Rayman. "Stand here," he pointed to a dark passageway to the garden, between the garage and the house. "We don't want anyone identifyin' you. I'll come out when I'm ready."

"He took out his FBI badge, and rang the doorbell. The badges were good, he reflected. His client's team hadn't screwed up those, at least. The right weight, but a little scuffed. Credible. And all had been delivered to his back-up crew in less than forty minutes, he reckoned.

Within a few seconds, he heard footsteps.

Who is it?" a young woman asked.

"Miss O'Reilly, this is Agent John Forman from the FBI. There's a sexual predator in the area, and we have good reason ta believe he's got you on his list of next targets. I've been assigned to protect you," he continued, holding up his badge four inches from the peephole.

The door opened a crack, it was still chained. A woman, he guessed to be in her early twenties, peered through the crack. She studied him.

"If ya don't mind, Miss, it's a little cold out heeya," Antoine said, smiling. *She's not buying it.* "It's tied to your story on pedophiles."

She unchained the door and let him in.

"I didn't think it would create that stir," she said. She turned and walked towards the next room.

In two steps, he took out his Taser stun gun and zapped her in the side of the neck.

She fell, unconscious.

Antoine flex-tied her hands behind her. He took out a cat toy tied into the thick rubber band and slipped it into her mouth, extending the band over and around her head.

Antoine proceeded to drag the journalist from the living room into the kitchen, which was dominated by a sturdy oak dining table. He lifted her up, and laid her on the edge of the table. Antoine removed the candlestick from the table and put it on the chair. He slid her to the middle of the table, unbuttoned her jeans and slid them off. He ripped her blouse off, revealing a bra, which he undid, making sure not to scrape any glove leather on the clasps. Antoine took three condoms out of his pocket and placed them on the table next to the young woman's body.

Antoine returned to the open front door. "Frankie, come on in, warm yourself up."

***

A half-mile away, Paul ran to the alley behind King's Stationery and Copy Center. He knocked on the back door. Nothing. He slammed the door with the palm of his hand. Paul waited another ten seconds before repeating, but harder.

"Please, oh please, come on," he whispered, not just in relation to this closed door, but to the whole situation. He had had no response to the SMS's he had sent by the time he had given his phone to Kendrey. None from Mazurski, Artois, Maggie – the people who would take him seriously.

And now Kendrey's mission at the police station was a big question mark.

He stepped five feet to the side of the building and looked – another thug sauntered past the alley. Paul pressed himself against the beat-up wooden garage door, hidden in the shadow. *No Time for this.*

As the employee opened the door, Paul approached.

The employee dropped the garbage bag he was carrying and assumed a Kung Fu stance. "Oh shit, mo-fo'. Hee-yah", the employee cried out, as he moved his hands in a badly-choreographed martial arts sequence.

"Whoa, Jackie," Paul whispered. He gestured for the copy clerk to take out his earphones.

The employee did so.

"I'm not the enemy of the people. Name's Robinson, reporter, WQHP."

"Wow, cool dude. Whazzup?"

"I need your help. Some murderers are cruising this town, and I need you to call the police."

Paul heard footsteps on the side street and pulled back against the garage door.

"Journalist O'Reilly's in trouble."

***

Watkins walked down the hall towards her, ogling her legs. *Too easy.*

She stopped and pored over the WQHP office directory on the wall. *That desk's not even in the News department.*

Let's see. . .found it. Tenth floor," she uttered quietly into the concealed microphone.

Watkins re-appeared in front of the desk he'd assigned her.

"Is the oil spill release ready?"

"Right here," the Temp replied as she held out two pages.

"Great," Watkins replied, smiling.

"Mister Watkins, I'm a little hungry. Mind if I step out for a quick hotdog?"

Watkins furled his brow, before replying, "I'm afraid that normally this is crunch time."

"Pretty please? You don't know how much I love those seven. . .inch. . . pups," the temp replied, with a pout.

Watkins opened his eyes wide..."Um, that's, that's. Oh, well, uh, o.k. if you're back in te-, fif- fifteen minutes," he stammered.

Amanda crossed her sweater with finger, tugging parts of it as if she wanted to take it off. "Cross my heart, and hope to..."

The associate producer looked up at her, while continuing to speak into his hands-free. "No, *fine*, take the Poconos background out. Put in the Liberty Bell. . ." He jerked his head up slightly.

"Hi, I'm a new admin assistant with the politics desk. I have a message I was supposed to deliver personally to Dr. Artois'. Where is she?"

He pointed to the right and held up three fingers. "Yeah, the third stock one, goldish-tint," he said briskly into the mic.

Amanda nodded and walked past two desks, stopping in front of third one. She looked back questioningly.

The producer brought his stare up from her legs and nodded.

She scoured Dr. Artois desk for information. She saw brochures on furniture, and another on home remodeling from Lowe's. She stepped to the computer on the desk. She glanced at a stethoscope as she hit the space bar of the keypad. The second password worked. The e-mails appeared. She vetted them. 'Emergency' was one subject, and she opened that e-mail. Again, she saw 'Stoltzfuss' and deleted that e-mail. Once again she permanently deleted the trash. She turned around and walked to the elevator.

***

Paul turned on his phone. Nothing. He started wondering why he wasn't getting calls. Actually, Artois' boyfriend David was the only one who knew the reason behind his being in Hollidayton. He would have told her, wouldn't he? Don't couples discuss these minutiae anymore? *Maybe he's not back from work yet.*

Paul turned off his phone again and removed the battery. He looked down the alley and watched the police car drive past on a side street, lights flashing, but no siren. *The message had gotten through.* He breathed a sigh of relief. Then he tightened up. *What if they're responding to a murder?*

Paul continued running through the alley, towards O'Reilly's house. One block more and he'd be leaving the business district and be much more exposed. With police patrolling, and the 'agents' still present, he'd have to be more careful for this last quarter-mile to the end of Lindsay Street.

He heard footsteps and pressed up against a warehouse door set in twelve inches from the alley. An 'agent' was patrolling along Spring Street, thirty yards away.

The shadow of the backlit church steeple cast itself over the door in which he cowered. Not detected.

Now he had to get over that fence and to the next street, without the help of alleyways.

The thug turned onto the next cross street and walked out of sight.

Paul sprinted across the road and started his leap over the stylized arrowheads on top of the waist-high wrought-iron fence with stylized. He didn't clear it properly and heard a loud rip. His left leg jerked backwards. He fell down on the thin patch of snow. With his right hand, he felt up and down his thigh. A six-inch tear.

A dog barked inside the house.

He got up and ran over the neatly trimmed lawn at a good chop. He glanced and saw someone pressing their face against the window. Paul dove behind the nearest tree, ten feet away, on the edge of the yard. A cool breeze caressed his rump as he waited thirty seconds for confirmation – had he been seen or not?

***

Artois greeted the four paramedics as soon as they arrived in the cafeteria. "It could be listeria, something sudden," she said to the chief paramedic, "I suspect they will only need liquids and bed rest, but the doctors should check for salmonella, too."

The paramedic nodded.

Upon returning to her cubicle, Artois tapped the space bar and the screen-saver disappeared. She tapped in her password. She opened her e-mailbox. The e-mail heading she had seen earlier was gone. She was sure it had been sent by Paul. "Emergency-Murder", the subject line had read. She dialed Paul's extension and waited five full rings before hanging up. She dialed Paul's number. Voicemail. She dialed Watkins' number.

"Hi, this is Doctor Artois-"

"The good doctor, so to what do I owe this pleasure?"

"Is Robinson there?"

"No, he isn't. You didn't get the news? He was let go yesterday."

"Yeah, I heard"

"Downsizing, downsizing, downsizing. It's the Flax News juggernaut."

"What passes for news nowadays. Listen, Jerry, before he went, did he mention anything about a murder?"

"No, not to me at least, but you know he's always horsing around. Any particular reason?"

"No, I just thought I received something from him with an 'Emergency' subject heading, followed by 'murder'.

"Didn't say anything to me. But you know how he is." Yeah, he's probably just being his impulsive, dramatic self."

"That's our man. I wish him luck. In any event, I'm looking forward to that special on asparagus."

"Thanks, Jerry, you won't be disappointed."

Artois hung up and scrolled again through her e-mails. She couldn't find any more from Paul, besides the confirmation of Sunday outing. She deleted it and returned to reading her script.

***

As he approached to within thirty feet of the house, Girelli glanced at his watch. It read 7:41.

From out of nowhere, a police car pulled up from the opposite direction, facing him, lights on.

Girelli had nowhere to go, nowhere to hide. He felt for the FBI badge and his gun. As per the instructions, he whipped out the FBI cap provided for the mission and put it on.

The car swung into a space in front of the garage, then pulled in straight until it stood parallel to the garage door.

Girelli walked confidently towards the police car. An officer got out, gun drawn. He stood a full six-foot-three, and broad-shouldered. "Stand right there, arms up," the officer commanded.

"Shhh, shhh," said Girelli, putting his finger to his lips, while holding an FBI badge high with his other hand. "FBI. Agent Jack Parducci," he said quietly, "We're conducting a raid. A dangerous sexual predator. Shut your lights off, please."

The officer stepped up and looked at the badge. Girelli stood still as the officer nodded, shifted the revolver to his left hand, walked back and switched the car lights off.

"Why haven't we been informed?" the officer asked.

Girelli took a step towards the car. "Because he's been tapping police department radio frequencies in every town he's operated in. Could you shut the radio off?"

The officer looked at his watch. "I can shut it off for four minutes."

"Fine," Girelli answered. "I just got word from my men across the street that he's in the kitchen, with her. If we act fast, we can save her. What's your name?" Girelli asked, taking out his gun.

"Warren Lee."

"Good, Warren. Now let me just call to see what they've picked up, on his location in the kitchen." Girelli pushed on his discrete walkie-talkie.

"Yeah?" Antoine answered.

"Chef, I'm standing outside the front door," Girelli started, in a hushed tone, "with an Officer Lee of the police department here. He was patrolling."

There was a long pause.

"Shit happens, but 'dis could work," Girelli finally heard, followed by instructions.

"Gotcha," Girelli answered. He turned to Lee. "He's already working on her. He'll be tense, but the good news is he'll be distracted . . . Warren Lee, you could be a hero. Nineteen months we've been looking for this guy." Girelli crouched down. So, how do we do this? Ya want me to go first?"

Officer Lee looked Girelli in the eye, his nostrils flared. "This is *my* town, I'll go first."

Girelli nodded, trying to emulate DeNiro in any of the movies where he finally gives his agreement. "Okay, Warren, this is your big chance," he said, focusing on the door. Let's see if it's open. Girelli tried to turn the handle. It didn't move. He turned to the officer, shaking his head.

"It's okay, I've kicked half-a-dozen of these in," Lee whispered.

He positioned his large frame in front of the door. He planted his left foot on the concrete landing and set his right foot in back. He pushed off the landing with the ball of that foot and propelled it just to the left of the lock. The wood around the lock splintered into twenty pieces, as the kick swung the door in. Lee burst through the door.

***

Antoine had anticipated the man would be big, so he had only to make a slight adjustment upward. The cop's surprised face was now three feet away from the silencer. Antoine pulled the trigger. A soft 'phtt'.

The officer fell like a stone, blood dribbling from his forehead.

Antoine bent down and removed the Smith & Wesson from the officer's hand. He returned to the kitchen, Girelli following behind him.

Tears streamed from the girl's eyes, as Rayman tore open the condom pack and started putting on the pink latex piece.

Antoine heard the door open and ran, crouched low to the wall. He swung his weapon around the wall, aiming at the door.

Stocky Fulton, from the second crew, walked in.

169

Antoine holstered his Beretta, but kept the Smith and Wesson out.

Fulton shut the cracked door behind him and proceeded to step into the mass of blood, skull bones and brain tissue that had fallen behind the officer's head.

"Oh Jeez, Fulton, watch where you're steppin' Now they're gonna get your shoe imprint. Haven't ya fuckin' seen C-S-I?"

The man froze in his place. "Sorry, Mr. Antoine, didn't see the blood."

"Didn't see? It's red, da carpet's beige. Didn't see." Antoine shook his head. "Now don't move. We're gonna get ya a towel." Antoine turned to Girelli. "Tony, bring him some kitchen towels."

Antoine returned to the kitchen and looked at Rayman, now draping the bare journalist's right leg over one of the dining chair backs.

25

Dupont glanced at his watch. Lee was overdue by two minutes. Dupont picked up the radio.

"Officer Lee, do you read?" Dupont asked.

The radio was off.

The Altoona police were patrolling the northwest perimeter of Hollidayton, but Lee was on the north side. Lee was following up on the call from the King Copy clerk about the journalist. Two cars were at the Panda Cafe, following up on an incident. *State police? Crazy men?*

An officer stumbled through the door. He sneezed. His complexion was the color of oatmeal.

"Jim, you're a big man for coming," Dupont started. "Any other day, I'd send you right back home to Linda, you know that."

Jim nodded, stuffing the handkerchief under his nose.

"But listen. You just man the phone here. Any calls, you tell me about, you know my frequency." He put on his police cap. "I'll be back after I check on Warren. It's been three minutes since he was supposed to call."

Jim nodded. "Sure, chief."

Dupont threw on his coat, grabbed his holster and gun.

"If I don't call back in fifteen minutes, send Jack and the Altoona cars to Spring Street. Better make that twenty. Those smart-asses'll never let me hear the end of it if I can't protect a journalist in my own precinct."

***

Antoine returned to the table. He opened the bag he had put down on the chair earlier, and took out a copper piano wire. He always used that in place of a garrote. It forced people to share the responsibility, building teamwork one small death at a time.

What he had to do now was, perhaps, unpleasant. But for the money, for his retirement package, it would be a small sacrifice.

171

And for the target, well, he recalled that crusaders had perished by the hundreds of thousands for their cause.

Rayman had just lifted the journalist and attempted to bring her hips to the edge of the table. She shook violently and he lost his grip. Her right leg slid off the chair back.

"Romeo," he said, turning to Rayman, "before you get started, you're gonna have to do something."

Rayman looked at Antoine.

"There are gonna be your fingerprints all ova' here. Dis ain't a problem, unless she identifies you," he started.

"No, no!" Rayman replied, shaking his head.

"Right. We ain't gonna let that happen, but I'm gonna need your help," he said, slipping the wire around the journalist's neck. "I'll need ya to take this wire in ya' right hand and pull. At least hold it with ya right hand. I'll pull from the other side."

O'Reilly suddenly started shaking her head and bucking. She lifted her legs and kicked Rayman away.

Antoine took his gun out of the holster, tossed it up in the air, grabbed it by the barrel and hit her in the temple. She went limp.

Girelli approached the table.

"Here, Girelli, take this end of the wire. Pull."

Antoine stepped away from the table, wondering how long it would take for the two guys to squeeze the life out of this young woman. Age helped, but only to a certain extent. Her dreams would soon enough be rendered useless and her accomplishments moot. The law of life when a high-paying client needed something. *Journalists are a plague anyway.* After a few minutes, he detected no more struggle. "She looks done."

He turned to Rayman. "Carry on, Frankie." He picked up the officer's Smith and Wesson; it felt a little crude in his hand, but it was a fine handgun. Technically, he knew he couldn't fault it.

What he had just done saddened him. But the next act would be more pleasant. Executing ne'er-do-wells, just like those sinners in the hand-painted reproduction of Hieronymus Bosch's "Garden of Earthly Delights" hanging above his fireplace, well that was almost a task he could look forward to, *un piacere* - a pleasure.

172

He stood now against the wall, a good distance to the right of Rayman. Unnoticeable.

Rayman once again lifted the journalist's right leg and draped it over one of the chair backs. He did the same with her left leg. He straightened up, in both senses, and dropped his hips back five inches.

Sick. Antoine lifted the Smith & Wesson and pulled the trigger. A deafening roar.

Rayman did not fall. He did not move. A clean red spot appeared, marking his right temple.

The cabinets and counter were now a spotty pink color.

The absolute silence was interrupted only by the sound of towel scraping wall-to-wall carpet in the adjacent room.

Antoine walked around to the other end of the table.

Rayman's eyes remained wide open, staring down at the journalist. It was a perfect lobotomy.

Antoine savored the sharp, biting smell of the cordite, as he slowly walked around the table. A good, perfectly-timed and placed shot, like a fine Barolo, one had to appreciate without rushing.

"Boss, in case you were wonderin', that's the way I wanna go . . . but wait 'til I'm finished," quipped Girelli.

"Don't get your hopes up, Girelli." He walked briskly back to the living room.

Standing in front of the dead officer he turned around and fired another shot into the doorframe leading to the bedroom, about chest height. He bent down and placed the gun in the dead officer's hand, making sure the palm 'cradled' the grip and finger touched the trigger. *He will have gone a hero, stopping a rape and shooting at a bad guy. Alas, he missed.*

He heard something hitting the table.

"She ain't dead, boss!" Girelli yelled.

Antoine quickly ran back.

The journalist had regained consciousness and was once again struggling against her constraints and jerking her head.

Antoine grabbed Rayman's limp hand; he crumpled the cooling fingers around the ring at the end of the copper wire. "Grab da other end, Girelli."

He then pulled hard, and felt Girelli doing the same.

"Where da hell is Franklin?!"

"Hey, Mr. Antoine, I need more towels. Maybe a big one, and wet!" Fulton yelled.

"Yeah, yeah!" Antoine dismissed the request as he pulled the wire. "Come on Girelli, pull harda', we ain't givin' her a massage for chrissake!"

"Sure, boss."

O'Reilly once again tried to move her hands around. Instead only her elbows moved from side to side.

Antoine pulled so hard on his end that both O'Reilly's head and Girelli's hand jerked to the left.

Girelli countered.

Within ten seconds of starting the grisly tug-of-war, Antoine saw blood filling her eyes. Another ten seconds and it was over. He let go and wiped the perspiration from his brow with the back of his glove.

***

Paul arrived at the cross-street and jumped over a two-foot hedge. *It has to be the second house from the end.* He saw the side of a police car, lights off. He heard a shot ring out. Then quiet.

No one running out of the house.

He jumped over the hedge and sprinted to the house. He stood at the corner, and quickly jerked his head around. No one. He approached the door.

An inch-wide beam of light issued from the doorknob. He edged his head towards that gap. He bent down, peering through the crack. The living room, a fireplace towards the right. On the mantelpiece were some music boxes. Light was emanating from the other parts of the house, which he couldn't see on account of the wall, but the kitchen had to be on the other side of the framed entry The kitchen was well-lit.

Suddenly something just behind the front door moved to the left.

Paul jerked backward. He waited five seconds, then moved back to the door. He looked down on a guy who was squatting, wiping something.

If only that baseball bat. He looked around, and noticed that the driver's door of the police car was ajar. From the story he'd done on police escorts for VIPs, he knew there'd be at least a 50% chance of a shotgun in the front.

He glanced inside the car. *I'm in luck.* He snapped off the brace and pulled the shotgun out. He cocked it.

Paul skipped to the door sideways. He lowered himself to the door and looked through the gap. The man was still squatting, but had now moved a foot to the right. He was about two feet behind the door, three inches to the right of the doorknob. The door would clear.

Paul leaned back and practiced the hit with the butt of the shotgun. He took a deep breath and lowered his left shoulder to within an inch of the door. He bent his knees. In one movement, he bumped the door and stumbled in, knocking the squatting man down. *Shit, didn't clear.* He regained his footing and pointed the shotgun into the face of the startled thug. Paul lifted his left finger to his mouth.

The thug's eyes flitted between Paul and the shotgun.

With his left hand, Paul motioned for him to lie down on the ground.

The thug lay face down. Next to him was a police officer, quiet, missing the back of of his head. Bits of light-colored matter lined the edge of the skull.

Jeez. Paul took a deep breath. *This guy'll be packing heat.* The question was – in a holster or back of his waistband? With his left hand, Paul reached down and flipped up the back of the thug's coat.

He stepped forward six inches, bent down, and patted his hand above the belt. A little to the right, he felt the hard contour of a hardened resin Glock handle. He eased the pistol out, then shoved it into his own back waistband. He felt another large drop of sweat falling from the end of his nose.

175

26

Five blocks after Dupont had left the station, the radio clucked to life. "The Navin's just called from 1003 Spring Street. They had heard a shot coming from across the street. They saw a police car, but wanted to confirm everything was okay. Over."

"Jim, have cars three and four back me up on Spring Street after they finish at Panda Cafe. Over."

Dupont cut the lights and motor as he coasted down the street. He stopped next to the last window. He quietly exited the car, crouched down, and walked towards the front side corner of the house. He glanced around the corner of the house, and saw the stock and butt of a shotgun disappear inside.

He jerked his head back and pondered his options. Never a good idea to go through the front door. The light was on in the kitchen, but not in the bedroom at the end next to the car.

***

As Paul pondered his next move, a draught of icy air sailed through, slamming the door shut behind him. He froze. He felt another drop of perspiration dangling from his nose.

He heard a faint huff and a puff from the dining room. Squeaking table legs, followed by a hard pounding on the table. Again, but not as strong.

"Fulton?" he heard a deep baritone ask.

Fulton twisted his head around and looked up at Paul.

Paul squinted his right eye, and jerked the barrel one inch closer to the thug's head.

"Yeah boss. Just checkin' for Ed. Still not here," hollered the prostrate thug.

Paul nodded. He stepped, right leg first, past 'Fulton.' Now standing in front of the man, he grabbed the shotgun's forestock with his left hand, steadying the barrel onto the center of the man's forehead. He proceeded another two small steps towards

176

the hallway, keeping his eyes and the barrel end focused on Fulton's well-chiseled jaw.

Fulton smiled. He turned his eyes away from the shotgun and towards the well-lit dining room. "Hey Frank, I need that other towel!" he shouted.

Paul felt chills running down his spine. He flared his nostrils and kept the shotgun aimed at the thug's forehead. *The creep was testing him.*

***

As Dupont passed under the window of the next room, he heard a table creaking. The action was taking place here.

Both the bedroom window and the storm window were slightly open. In this neighborhood, no one needed to close their doors and windows . . . *usually.* He reached onto the edges of the storm window, pulled in each of the releases and lifted it higher up. He pushed the window farther open and hoisted himself up to the window. With both elbows resting on the window sill, he quickly brushed open the curtains. No one. It was silent. Dupont slithered through the window and dropped, quietly onto the thick carpet.

"Now go get a towel from the master bathroom. And make it wet," he heard.

Dupont quickly pressed himself against the walk-in closet door, backing onto the room where something was happening.

Three seconds later, the bedroom door swung open with a loud squeak. *Don't look to the right.*

A man in gloves and black suit walked briskly past the closet and into the bathroom. The bathroom light went on. The man would see him as he got out, so Dupont took three quiet steps to the bathroom.

The man exited, holding a towel.

Dupont pointed the gun between the man's wide-open eyes. With his other hand he held his finger to his lips.

"FBI," the man said meekly, dropping the towel.

"Reach for the badge slowly, with your right hand," murmured Dupont in a low voice.

177

"Make sure dat thing is big n' fluffy and wet! We got brains all over da floor!" Dupont heard from someone in the dining room.

Dupont lifted his left eyebrow. "Go back to the bathroom," he whispered, "and turn on the water."

The stocky man stepped backwards towards the bathroom door, stepped sideways, and slowly, with deliberate motion, opened the faucet.

***

Paul edged his way backwards to the wall, still keeping the shotgun fixed on the guy's head. He felt the wall at his back. The fireplace was to his left. He slid along the wall about a foot, until he felt his ribs touching the wooden molding around the passageway to the hall.

As quickly as he could, he slid his left hand down the forestock, past the trigger guard and up to the stock. He brought his right hand up to the forestock and jammed the butt of the gun up against his left shoulder. He'd shot a rifle southpaw-style lots of times in his youth, but fifteen years later, it felt odd. No choice here, though. If he shot from his right side, he wasn't sure of getting the barrel end to clear the wall. Besides, his body would be leading and therefore providing a plum target for the thug in the next room long before the shotgun barrel ever appeared.

Once again, he felt his left hand throbbing as he lightly fingered the trigger. The drops sliding off his nose had turned to rivulets.

Slowly, he turned his body towards the fireplace, still lining up the shotgun sights on Fulton's nose. His hamstrings were straining as he bent his knees. He was now set. *Jump backwards against the wall. Pivot left. Shoot someone in the middle of the kitchen.* He repeated the plan twice in his head. He lifted the shotgun barrel up and held it parallel to the wall. He crouched slightly, balanced on the balls of his feet. Finally, he turned his head straight and faced the side of the fireplace mantel.

"Boss, lookout!"

Paul whipped his head back to the right.

Fulton had started pulling out another pistol from a shoulder holster. Paul swiveled the shotgun back down. He fired. The thug's right eye and ear disintegrated.

But the recoil bounced Paul's shoulder against the wall. He lost his balance completely. Falling backwards into the hallway, he let go of the shotgun.

He was sprawled out, his upper torso and head blocking the hallway. Only his lower legs were obscured by the wall. He sat up.

A large hand, holding a gun with a silencer, appeared out of nowhere.

Paul sat still. He recognized the man. It was the boss.

The man glanced to his left. He didn't need more than a second to see his henchman was no longer among the living. He stepped forward towards Paul, moved the barrel of his gun to his left and pulled the trigger.

The bullet tore into Paul's right leg just below the knee. He'd had never felt such pain before. "AARGH. Shit," he jerked up grabbed his leg. Blood oozed between his fingers.

The boss, the one he had seen in the Explorer, strafing his car, had just crippled him.

That psycho now calmly stepped around the wall and stood next to his right hip.

Paul had barely recovered from the shot in the leg, the psycho kicked him in the left rib. "Aahh!" Paul screamed, his upper torso falling back down.

The man stepped over Paul, and kicked the shotgun away.

"You stubborn, curious shit. Don't even try to run. Da journalist is dead. Ya' finished."

Paul held his right leg, grimacing in pain. *No mention of Kendrey. They don't have him yet.*

"Now kid, you were about the toughest mark I've eva' had to score. Two . . ." he started, looking at Fulton's body. "Three good men dead cause a' you." He lowered his head and shook it. "In otha' words, a good gladiator. And the good gladiators, when it was time to go, they had a choice 'a death. Ya know, for bravery, the lion or the leopard." The strange-sounding man in the black coat stared down at Paul. "So . . . do I shoot you in the kidneys . . . or the liver?"

Italian. Paul observed the scar on the man's forehead and the one on his cheek. He couldn't count on mercy. He couldn't reach for his new pistol either. Could he try something else? "I need both—I drink lots of Molson's."

The Italian didn't smile. "A comedian too. Dat's cute." He lifted the barrel effortlessly.

Paul saw a little puff from the gun's silencer, and felt a hot prick on his left ear. "Ow, ow! Jeez!" He grabbed his ear with his left hand. His fingers felt warm. His eardrum was ringing He brought his hand down and his fingers were covered in blood.

"Now listen up good, with ya' other ear."

The electronic melody of "O Sole Mio" went off in the Italian's pocket. The guy deftly took the pistol in his left hand and pulled the phone out with his right one. "Yeah?" he answered, his eyes focused on Paul.

"Fine. We got her. Rayman is dead too. Yeah. Set up just like you wanted it." He put his phone away.

"So where were we? Oh yeah, anatomy. Ya see, the liver's down there, a little to the right and up from your belt buckle," the Italian continued, coolly.

Suddenly, from the master bathroom, the sound of glass shattering.

"Cop!" someone yelled. The scuffle continued beyond the half-closed bedroom door some twenty feet away.

The Italian kept looking at Paul with his beady eyes, but his mind was somewhere else, calculating. After three seconds, he turned back to him. "Lie down, chimp-reporter!"

So THIS is the reputation I'm taking to my grave? he clenched his teeth.

The boss ran towards the table, picked up Girelli's Glock lying there, ejected the magazine and the loaded round in a millisecond. He took two steps towards Paul and threw the gun on his lap.

"Pick it up and aim it at the bedroom door." The Italian pointed his silencer at Paul's forehead.

Paul picked up the pistol and stretched out his left arm. He lowered himself onto his left side. Pain shot up from his right leg. He grimaced. Slowly, he pushed his right hand against the floor, lifting his upper body. He rested it on his left elbow and aimed the barrel in the direction of the bedroom door.

Another glass broke against something.

Paul focused on the Italian. *I will end your days on earth, I promise.*

"Look over there to da bedroom. Ya look somewhere else and I'll drill ya' fuckin' skull," the Italian muttered.

Paul returned his gaze towards that door. As his eyes swept the room, he had a glimpse of the couple on the table. They were frozen, the man's gluteals tensed. Her legs were on either side of his.

More commotion from the bathroom.

More pain in his leg.

The Italian took a step and dropped to his knees, then lay down on his right side, his body at a 45 degree angle to the hallway wall, his legs splayed, with his right foot a yard from Paul's head. The man's head was under the table, just behind the standing stiff's foot.

Smart S.O.B. To someone looking on from the bedroom, he would appear dead.

The guy jerked his head up and glanced at Paul. "You move and ya' finished."

It sounded like someone's head hit a mirror. "Aarghh!" Paul heard through the space between the door and the doorframe. *Who's luck will run out first?*

His gaze affixed on the bedroom door, Paul slowly reached back with his right hand. Any sudden movements and the Italian would know something was amiss . . . and he'd be dead.

He now held the deceased thug's Glock gun handle precariously between his thumb and index finger. He couldn't just grab it, since the handle was facing left. Any sound of gun falling and he understood psycho-boss would drill him in the forehead in a second flat; this guy was that good. He'd seen one like him once at the range. Been handling weapons since he was five. A 'handgun Mozart.'

Paul slid two inches of the barrel out of his waistband.

No sound from the Italian.

Another two inches of barrel out'.

The Italian jerked his head up again.

181

Paul froze his right hand, while continually staring at the bedroom door. He sensed the guy examining him.

The Italian lowered his head back down.

The muscles in Paul's left upper arm now also joined the burning chorus of other muscles. The combination of pain in both biceps finally overwhelmed him, so he lowered his left hand, and the empty pistol just dropped to the vinyl floor.

The Italian jerked up. "Gun up, asshole!" he muttered through his teeth.

27

Paul raised the useless gun back up. He gritted his teeth. He waited three more seconds, then, hearing no sounds from the guy, continued pulling the Glock out of his waistband. Five seconds later it was free and clear. Glacially, an inch at a time, he lowered it onto the carpet next to his left buttock.

He heard the bathroom door creaking.

His gaze still affixed on the bedroom door, Paul rotated the Glock so he could pick it up and fire it fast. He brought his hand slowly across the cool metal of the barrel.

From the direction of the bedroom . . . silence.

Paul slid his right thumb under the left side of the grip, and now clenched the grip firmly in his entire hand. He slipped his right index finger in front of the double trigger.

He heard the Italian's coat rustle as his head jerked up, then lowered again. *Now?*

Suddenly the bedroom door opened all the way. A short man in a police uniform, nose bloodied, stood with legs shoulder width apart. He was gripping his pistol with both hands, aiming first at the dead couple, then at the Italian, then at him. The man pointed the barrel straight at his head.

Paul let his left hand drop. *God, don't shoot.* Instinctively he closed his eyes.

"PPhht" he heard.

Paul opened his eyes.

The policeman grasped the doorframe with his left hand, a new red dot on his neck. He shot blindly somewhere between him and the Italian as he crumpled, slowly. "Pphut." Another red dot appeared, this time on his forehead.

Now!

Paul whipped the Glock from under his rump, took aim at the Italian's back, and pulled the trigger. A loud crack. The coat jerked forwards. Two seconds of silence were broken when the Italian lifted his torso slightly, whipped his Beretta up and shot back blind, hitting the wall six inches to the left of Paul's head, sending plasterboard flying.

Did I miss? Paul fired again, lower. He hit the Italian either in the butt or his left kidney. Polyfill from the coat, mixed with red, splattered onto the table leg.

The Italian jerked his head up, firing twice-in-a-split-second, then dropped his head.

Paul felt the bullet passing into his left side, hitting one of his ribs with sledgehammer impact. "Aargh." He was overcome with nausea. His head started spinning. He shot back blindly, hatefully at the thug boss. Once, twice.

The Italian didn't fire a fourth time. *A chance.*

Paul took a deep breath, then another. He aimed the gun just above the Italian's right shoulder joint and fired. He held his breath. Ten seconds passed like ten minutes. The muscles in his forearm burned mercilessly. He held the gun steady.

The Italian's head fell to the ground with a thud, his muscles loose. *He's gone.*

Suddenly everything started spinning. He waited, taking two deep breaths, then put down the gun. He pushed on the ground with his arms. His body moved back three inches. A searing pain shot through his left side. He repeated the movement until he was leaning against the wall. The room spun around.

He regained consciousness in what he thought was a few seconds later, but he didn't know. He looked straight ahead, beyond his bloody legs, beyond the black coat, torn and glistening with blood.

The table was an old mahogany piece, probably one that O'Reilly had inherited from her mother. In front, an erotic still-life. Paul looked up at the nude man's closely cut hair. In spite of the carnage, the silence of death, the lights continued to glow, as if four friends had just sat down to play a game of Monopoly. 'What if board games were also played to the death?' he pondered blindly, trying to ignore the rivulets of pain coursing through his body.

Out of the corner of his eye, Paul saw the front door open. He whipped his head to the right and flailed with his arm, gun at the end.

184

It was Kendrey.

A flood of relief. He smiled and leaned his head back against the wall. "Josh, Josh, Josh. Good thing you're here. Goddamn, I need an ambulance."

"Where's the journalist?" Kendrey asked, coolly.

"Right there on the table," Paul gestured with his jaw. "We're late."

"What?"

"She was killed by someone in this room."

Kendrey stepped over Paul's legs. He started examining the pair from every angle. He crouched and studied the Italian.

"Terrible," he said, shaking his head, "looks like. . . " he started, his jaw locking, "too many of the wrong people . . . " he continued, once again his jaw locking, "have guns."

Kendrey continued on to the next room. He bent down next to the policeman.

"Come on, Josh, I know you wanna look around, but I'm bleeding here. Call an ambulance, will ya?"

Kendrey nodded. He stepped back over Paul and glanced at Fulton's body. Kendrey had the same confused look he had when Paul told him he'd be taking a road trip.

"Where's the phone?" Kendrey asked as he bent down above Fulton in the living room.

"I dunno," Paul muttered, irritated. "Somewhere in the kitchen, probably!"

Kendrey stepped over his legs one more time.

Paul looked back to the left side.

Kendrey stood ten feet from him, now holding the henchman's gun in his right gloved hand. He dropped to his knees to examine something more closely.

"Will you call the goddamn ambulance, Josh?"

Josh got up and dialed three buttons on the stick phone. "Ambulance, to 1138 Spring Street. Five bodies on the ground, some wounded." He paused. "Yes, please in-inform the police."

"Thanks man, you're a hero. You know, you asked me in the car, Josh, how I got started in this.". He felt a pain in his left side, as if the local football team's kicker had been practicing all afternoon on that one spot. He sucked the air in shallow breaths,

185

through his front teeth. "Well, when I was growing up, I'd watch the six o'clock news every evening with my family. Dad, Ma, my brother, sister, Grandpa Jim. Grandpa Jim would let me sit on his lap and you know, burp in my ear," Paul started, looking up.

Kendrey put down the phone and picked up a burgundy-colored laptop from the kitchen counter, but kept the gun in his other hand.

"The house could've been burning down, baby sis' choking to death on jerky, and I swear, it didn't matter. We were watching Tom Brokaw. Well," Paul continued quietly, "by the time I was eight, I knew what I wanted to be. And when I had my first job in Harrisburg, a year after Brokaw left, Dad said, 'When you go on the boob tube, son, make it count. Don't just be another head talkin' gibberish and feel-good manure. Be like Brokaw.' And whaddya know, Josh, what's been my most memorable story, the story that defines my legacy?" He contorted his mouth into a chimpanzee's grin and screeched.

Kendrey nodded.

Paul heard a sound to his left.

One of the black-clad henchmen opened the bedroom door, his face a criss-cross of scarlet lines with mirror shards in sections.

Paul tightened his grip on the gun. "Look-out Kendrey, thug at nine-o'clock!" He whipped his arm up, shooting the man in the arm or shoulder, he couldn't tell. But the man went down.

A few seconds later the front door opened. He blindly he lifted the cocked pistol.

From somewhere, two explosive gunshots rang out.

Paul felt a sharp pain below his neck. The room began spinning. All was dark.

28

Occasionally, the gray clouds turned menacing and let go a barge-full of rain. It was the afternoon of the third day, and the CSI team was completing the evidence-gathering at the house. So far, Lieutenant Olson had only the following anecdotal information: the pedophiles, between the two of them, had fired sixteen rounds and one shotgun shell from their four guns. One pistol, strangely with a silencer, had even appeared to pass hands between them, and one firearm – a shotgun, was used to settle an internal matter.

Olson, now with temples graying, had worked closely with the CSI of the Pennsylvania State Police, as the Troopers were officially called, for twenty-six years. And he'd seen a lot. But the many years of experience under his belt hadn't rendered the gruff, stocky man emotionally immune to a bloodbath on this scale. Five, maybe six dead. Three, perhaps four more, gravely wounded. Amongst the dead, a police captain, one of his officers and a young journalist, who had been 'prepped' for a sexual assault and then garroted to death. He turned to John Finnegan, his assistant with the even plumper face of the Pillsbury dough boy. "Could it be that they couldn't agree on who would be next?"

"Could be. But what really gets me, is how Officer Lee managed to shoot Rayman in the head like that. The angle was impossible, unless Rayman was turning his head."

But the splanner pattern would be there," Olson pointed to a spot above the toaster, "not where it was."

"Unless he shot him and then retreated, at which point he was shot by someone else. But when did he have time to do that, and fire the shotgun?"

"Someone else had the shotgun, that's clear."

"Yeah, they had a whole gang, judging by the gunfight that took place outside, too."

Olson shook his head. He then summarized how some of the bullets had been found, three of them still in the bodies, eight through-and-through's, one grazing the skin of some other participant and lodged in the wall. The slugs would all have to be examined. It would take weeks.

"What's sure is that there were the four corpses here and another—judging by the brain matter scraped off the floor—missing." Also one instigator, perhaps the kingpin, lying in a coma in the hospital. Something Hollidayton hadn't seen since its founding and wouldn't see again until its proud steeples sank back into this rich Central Penn soil.

"And what about," the straight-laced Finnegan replied, "the three persons who aren't accounted for. Blood from two other persons on different surfaces as well, including the bathroom."

Olson said nothing, but nodded with his square lantern-jawed face, now rounded from years of eating solid Amish-inspired cooking. He walked once again between the doorframe and the point in front of the wall, where blood splatter vaguely framed a body of a certain Paul Robinson.

"The second shot was almost impossible, unless the chief lunged forward."

Detective Fred Agnelli emerged from the bedroom, with three separate evidence pouches.

Finnegan was still scratching his head. "Why did they target her?"

"Ya never know with these creeps," Agnelli chipped in.

"Another thing I don't get, boss. Why did they travel all the way from Harrisburg to get her?" Finnegan continued.

Agnelli turned to Olson. "We haven't identified any of the girls in the reporter's computer, either, but two will get you ten that they were from all over South Central Penn."

Olson grimaced. "Now that's about enough speculation for this afternoon, boys." He scratched his head. "I can't figure those perverts out—and it looks like you guys can't either, so let's just let the evidence speak for them. We'll know more in a week. Just make sure they've gathered every single atom of information at the scene."

The local and Harrisburg press hadn't had time to fit the shoot-out in Friday's journal. But on Friday, there was a full page photo of O'Reilly's house in the paper, front door sealed off with yellow tape.

Later that day, Olson surveyed the remains of the Stoltzfuss farm for the first time. They found the two bodies near the dead buck. One shot, the other with no noticeable markings.

Olson sighed as the last snowflakes melted. He turned to Agnelli. "It would be like looking for a needle in the haystack . . . not knowing if the needle was there in the first place.

"Yeah, chief and with shoeprints, forget it. The best starting point'll be ballistics."

Olson nodded. Whichever bullet killed the guy would be traceable to a rifle and subsequently to some hunter who, invariably, had panicked and run off.

The CSI Lieutenant grappled with the logic of someone running off. The empty chamber in the revolver meant they might find some blood linking the incident to foul play. Perhaps someone had been struck by the bullet. But what he instinctively felt, what he was sure of to the very marrow of his bones, was that this incident was related to the two pedophiles in Hollidayton. It had been a bad moon rising kind of night.

He had much to work with. Officers of the Harrisburg police had picked through Robinson's previously ransacked apartment. Fortunately, Olson reflected, the Harrisburg police had only gummed up the evidence a little. His trained officers from the State Troopers were on the scene before they had messed things up royally.

And one of the first things his people had found was Robinson's laptop, prominent on top of the mess on the floor.

Within a few minutes of receiving the portable computer at the lab, the techies found files containing the photos. Robinson, with young teens and, in some cases even younger girls. In a few cases, he was molesting them. Other photos revealed him together with Rayman and other female minors. Some of the minors were blindfolded and tied up. All were nude. The police found one-thousand eighteen photos in total. "Sure as hell didn't get there 'by accident' when trolling for legal porn," Finnegan said.

189

That Friday, the news had filtered to the major city journals as well, in the form of a nationwide story. Olson noted that the angle was a little different; the focus was on a "Reporter from leading broadcasting company," operating a pedophile ring with the apparent support and participation of a paroled pedophile called Frank Rayman.

***

Captain Olson stared at Latanya. "You say you had no clue, no indication that he was a pedophile?"

"None at all. I still, I still . . . " Latanya couldn't stem the tears streaming down her face. Latanya stared at the *Philadelphia Inquirer* cover story, with a shot of the house where the evil deeds took place. Next to it, two photos. The first was of some tanned body-builder-type, smiling next to a couple of kids. The second was of Paul, holding a chimpanzee on his lap. She seethed as she read the line again—"Seemingly innocent and well-liked reporter, had more disturbing child pornography on his hard drive, than had previously been seen anywhere in the region. A correspondence had taken place over the previous five weeks . . . "

How could he?!

"Did he disappear on you for days at a time, maybe to a hide out somewhere?"

She shook her head. "Oh sure, he was going somewhere for a story all the time. In fact, he'd even gone to Dayton a couple of days back. But he always came back with a story." She continued with three or four examples. "Have you talked with WQHP?"

"Yeah we did. But back to you – could you be sure that's all he was doing?" Olson asked. "Many a pedophile – and serial killer for that matter - had been the patriarch of an upstanding, conservative family, while committing atrocities."

"Well," Latanya reflected. She scratched her head. "I guess. I can't be 100 percent sure. He could, I suppose, somehow have hidden this . . . but, but, how?" She pushed the thought away, forcing back tears.

"Pedophiles are creative, resourceful. If you think about it, they have to be." Olson got up and walked five steps. "Have there

190

been times he shut off his computer as you walked in the room. Have you surprised him, somehow, to the point he reacted violently?"

Latanya thought of Paul, now lying in a coma in the hospital. She couldn't remember any such time, except once, when he was writing her an e-birthday card. She was surprised by his anger . . . but it was all an act. The card had been quite sweet. She wished he was awake, so he could talk to her, make sense of these accusations. She shook her head. "No, Officer Olson." She just wanted this to stop.

"Was he oddly close to children?"

Again, she had never noticed anything out of the ordinary. In no way could she corroborate the ' . . . evidently luring children with his charm and humor' one newspaper had suggested. "No. He kidded around with his nephews and with kids on the set, but nothing you know . . . *weird*."

"Fine," said Olson. "That's all for now. Call us if you think of anything." He handed her his card.

Latanya nodded, attempted a weak smile, as she took the bone-tinted card. She looked at it, then looked up at him. "Could I see Paul?"

Lieutenant Olson shook his head. "Related kin only."

"His mother can't leave home. His father is sick."

Olson scratched the back of his head. "I'm going in an hour. Meet me at the front of the hospital at three-fifteen."

"Thanks, lieutenant."

"Alone."

"Sure," Latanya murmured.

***

As she exited the interrogation room, Agnelli turned to Olson. "Why'd you do that?"

"Cooperation. I suspect we'll need hers to get to the bottom of this."

"Sure. I get it. She'll be on our side if we give her the small stuff."

Olson nodded. "Not only. Check her out, too. Who knows?" he said as he picked up the case file. "She could've been an accomplice."

"Should we ask for her computer?"

Olson nodded. "But let me ask for it. If it's clean, then we'll come back to her when he wakes up."

"And if it's not clean?"

"We are an equal opportunity incarcerating institution," he said, shrugging his shoulders. "But honestly, Finnegan, I hope not. As it is, after all I've seen over the past two decades, I have mighty little faith remaining in humanity."

29

Latanya got in her Honda Civic and drove off. She drove through the streets of Harrisburg, not going anywhere in particular. She passed the vast Agriculture Expo Center, ignoring it entirely.

Latanya felt something amiss. On the one hand, she'd been blindsided. They claimed to have photos. Proof. On the other hand, she knew what Paul sounded like, felt like, and even smelled like, when he was going to "conquer" a story. And he always did. This time was no different. His single-mindedness and sense of mission stuck out; she loved that in him. How could he reconcile that with underage girls? What about his passion for her, the love? Wasn't that it, the way he looked in her eyes . . . laughed at her casual remarks. His gentle touch. *Wasn't that love, plain and simple?*

She pulled the car in front of her apartment building. She couldn't hold back any longer, unleashing a torrent of tears. "Oh, God, why me?!" she repeated three times. "Damn, damn you, Paul Robinson." She grabbed a handful of tissues from the tissue box. She didn't want to go to the hospital. But she wanted to call that round-faced lieutenant even less. She went upstairs to shower and look presentable once again. But she felt like putting on one of mom's grayest, ugliest, most distressed mu-mus.

On her way to the hospital, she stopped in front of a kiosk and picked up six different papers. She arrived in the visitor's parking lot ten minutes early and started reading.

She looked closely at the picture of the wiry man in the newspaper. She had never seen him before, and Paul had introduced her to all his friends and acquaintances, *hadn't he?* Paul needed his time alone, sure, but did he have the cold-bloodedness to carry on a professional and secret relationship with a . . . child molester? To partake in those heinous acts, and host a wonderful dinner party . . . to caress her softly afterwards? Was he a Hannibal Lecter?

No! This was a set-up. She lifted her right middle finger to her mouth and chewed. *By whom, wasn't clear, but she would find out how it was done.*

***

Olson stood to the side as the female officer emerged from the room next to Paul's. Latanya ducked inside Paul's room. He appeared so helpless, all bandaged up. A dozen tubes running out of his legs, his chest, his nose. She shook her head as she watched the ventilator pump up, then down, simulating his breathing.

"It was apparently a miracle he survived. Another few minutes and he'd have been a goner. The doctor said something about a bone chip against an artery to the head. Lodged then dislodged. We found him on his side, in this same state," Olson gestured towards him.

Latanya stared silently.

"He was lucky. I don't know if society at large is," Olson continued.

Latanya turned and glared at Olson.

He didn't meet her stare, but must have sensed the anger. He shrugged his shoulders. "Perverts and killers come in all shapes and sizes, Miss Forrester. Someone in Hollidayton dropped off a laptop they had found in the bushes yesterday. It was the journalist's. She had found one of those pictures too. With Robinson."

Latanya once again looked down at Paul, connected to a respirator and a monitor. The prognosis for his recovery was not good—less than fifty percent. She shook her head then wiped away a tear. *No way, are you guilty. Something is off.* But this lieutenant wasn't, she felt, going to try very hard to find what that something was.

Olson turned to face her squarely.

She smelled a whiff of cologne, something her dad might wear when accompanying her mom to a social function, which he did rarely.

"Miss Forrester, I'd like to clear you, and to do so, we'll need your computer."

She looked at him, trying to hide her disbelief. "My laptop?"

"Standard procedure in these cases. You understand, as the girlfriend of the accused, you—"

"Fine, whatever," she said, reaching down to her bag and taking out the notebook device. "How long do you need it for?"

"Two days. And any flash drives you might have."

"Don't you need a warrant for that?"

"Do I?" he said, cocking his head.

"Forget it." She shook her head and reached into the bag, pulling two out and handing them to him. "The rest are in my office at State. I'll bring them in tomorrow."

As she walked to her car, she realized the first thing she had to do was connect with someone who could set her straight on the question of the photos. How did they find them? Were they on his e-mails or solely on the computer?

She had participated in a symposium on the characteristics of deviant behavior in adolescents the Forensic Science department had organized the previous Spring. On her desktop she looked up the e-mail of the professor that had recruited her to participate. Eventually, she found the number, picked up her Smartphone and dialed four digits before aborting the call. Latanya looked up at the ceiling. "I guess it's gonna be just like in the movies now," she murmured.

She drove to the nearby electronics store and bought a disposable cell phone. She paid in cash. After uploading 24 hours of calling time into the phone, she dialed the number.

"Hello, Jack McKinley here."

"Hi, Professor McKinley, this is Latanya Forrester calling. Do you remember me?"

"How could I forget. Great insights on the social stigma of autism."

"Thanks, professor. Now it's my turn to ask you to participate in a discussion, of sorts."

"Okay."

"I recall you mentioning that many aspects of a criminal's activity on the computer can be traced."

"Yes, it's true."

195

"Like sites he had visited, documents viewed,"

"That's right."

"Well, I need to know if my boyfriend had opened certain images on his computer."

"Hmm. Now, first off, you should be aware there could be legal repercussions if he finds out you did it."

"I'll risk that possibility."

"Fine. Can you bring in his laptop on Wednesday?"

"Wednesday . . . Um, I dunno," she scratched her head. "Is there another, uh, possibility?"

"So you don't have his laptop? Wait, you're not . . . this isn't linked to Hollidayton, is it?"

Latanya looked up, pondering what to say. "Can't hide anything from a forensics professor, can I?"

"Not really," he replied in a deadpan voice. "Why don't you come see me tomorrow, around three."

As she hung up, she wondered what she had just committed to. The laptop was lying, no doubt, on the desk of some investigator.

But she had one more contact, one more lever to pull. She pulled out her phone and dialed a number.

Two seconds later, the line was picked up. "Cap'n Tubman," her deep-voiced uncle answered brusquely. Latanya hesitated. It seemed that whenever she called this number, she had cause to be embarrassed, whether it was helping her brother Michael beat charges for drug running, or protecting some classmates from being declared a public nuisance.

"Hi, Uncle Charlie, it's me."

"Why hello, Latanya. Wait, let me guess . . . it's Michael again."

"Um, no, Uncle Charlie, as hard as it might be to believe, no, it's not."

"Well, well, this is the dawning of a new day."

Latanya smiled.

"It's something about . . . well, I'll need some advice. Um, could we meet today . . . and maybe not at the precinct, it's sensitive."

"Hmm, Latanya, sounds intriguing. I've got a meeting at two. How's half past three, at the Village Coffee Shop."

"Fine. Thanks, Uncle Charlie."

"Jes' don't tell me it's your brother again."

"No, swear to God it isn't."

***

Agnelli moved to another angle and studied the situation. He set up the laser pointers on tripods and the table. They looked at the photo taken of Rayman. Each of Rayman's hands held two handguns. Because the bullet that penetrated his skull lodged in the frontal lobe of his brain, his body had lost the capacity for movement.

According to the angle, the officer had moved to the right, then was moved back to the left by what? The shot from Rayman's firearm? Didn't make sense, as his muscles would have stiffened immediately; he had been transformed into a human statue instantly. Holding two handguns and sporting an erection. He definitely could not have moved once he had been shot. Olson smelled a set-up.

"The angle of the shot was strange," Agnelli blurted out, "either officer Lee was somewhere else when he made that shot . . . or else the reporter had been getting up. The question was— what caused Rayman to shoot his partner in the leg and side? Accident?" He shook his head and smiled, in frustration. "Anger at Coitus Non-Startus?"

Finnegan didn't react. "Another thing I don't get is, uh, how could this guy have shot two police officers, his partner, and a third, unidentified victim, while preparin' to do the journo."

Olson didn't reply. He had his own hypothesis to focus on. "There was blood from two other persons who were no longer there. Some brain matter from one of them, too, so it was a corpse removed from the scene. And another who left some blood on the master bedroom bathroom sink?" Olson shook his head.

Finnegan shrugged his shoulders.

197

"And, let's see, there was blood from two other persons who were no longer there. Some brain matter from one of them, too, so at least one corpse removed from the scene. And another who left some blood on the master bedroom bathroom sink," Agnelli continued.

Finally, they had managed to get away in the police car and two black SUV's . . . unnoticed?" Olson asked.

"Well, the rest of the officers had been patrolling the other part of town. The captain hadn't radioed in for support," Finnegan said.

Within the span of six or seven hours, at least six people, including two police officers, had been killed or died under inexplicable circumstances, and, judging by the blood loss in other spots where gunfights had taken place, on Oregon road, for example, the death toll was even higher.

Furthermore, the biggest unresolved mystery loomed, like a charging elephant. *Another man was here and appeared to have been ambushed.*

"And most of these casualties caused by a child rapist ex-con and a reporter?" Olson shook his head. He was intrigued by the pedophilia charge. He hadn't yet seen any of the photos, but Finnegan had, and his only comment was that they were "depraved." "No hypotheses make any sense here, guys. None. Zero."

And furthermore, time was not on his side. He knew he could put to bed those myriad possibilities only after his lab had reconstructed the scene as it had taken place. And whatever scenarios ended up being possible, they would all have to factor in three blood sets belonging to persons not found. Eyewitnesses kept saying 'State Police', and one of the Navins claim they saw an FBI cap on a man talking to Officer Lee. But he had followed up and the FBI only had Rayman in their NSOR or National Sex Offenders Registry database, and their point man said clearly they hadn't been after him *for years.*

Finally, a crazy man was thrown into the mix, according to the Café Panda witnesses.

198

Only one ex-con came up on the CODIS matches, and he was the one missing about a quarter of his brain.

"I don't think even a low-level scumbag like Rayman could get by with that much missing upstairs, but who knows?" Finnegan commented.

Olson ignored the black humor. "Harry, so you tell me that it's possible Captain Dupont hit Robinson in the chest while Robinson was bending over. But from that angle, Robinson would have been propelled backwards, towards the front door, not to the wall, which would have been to his left . . . " he paused, awaiting feedback.

He only got a nod and shrugging shoulders.

"Yeah, Harry, it's a lot of bullshit. Robinson's a country boy with a hankering for gun ranges, not an Olympic speed shooter. The only fingerprints we find are on the shotgun . . . and partials on an Uzi at the Stoltzfusses. The story falls apart starting with some guy on the ground next to Rayman . . . hit at least twice by bullets from the reporter's gun." Finnegan continued.

"Finnegan, put out an all-state bulletin with the faces of the—" he inverted his fingers to signal parentheses "State Policemen."

30

The Village Coffee Shop was a noteworthy establishment primarily for the fact that it hadn't changed in thirty years. Same Formica tabletops and, it seemed, cardboard ice cream ads.

Chief Charles Tubman was an icon of the force. Harrisburg had had its ups and downs, and he had charged forward, meting out justice on the streets, while also making sure that the bad guys were separated from the n'er-do-wells. He saw to it that the latter weren't just thrown into jails as collateral damage. Already, life for the average African-American here was no picnic. Latanya knew that as well as anyone. But still, she remembered Uncle Charlie telling her that "At the end of the day, you're judged on what you do, not on some sorry-ass excuses. Life ain't easy for anyone but the Rockefellers."

So far, her life had proven him right.

He strolled in as if he owned the place. His square jaw, now rimmed with gray hairs, still commanded respect, both inside and outside of the force. When he hugged her, almost all the air flew out of her lungs.

They settled into a table in the corner.

After the questions on how her dissertation was moving along and the family, he was ready to listen.

"Uncle Charlie, here's the situation. My boyfriend is the reporter in the hospital, the one accused of being in the pedophile ring."

Her uncle suddenly looked down at his cup of coffee and shook his head.

"No, listen, Uncle Charlie, I know him, he didn't do it." She explained what the forensics professor had told her. "But I need that computer. The laptop."

"Honey, I can see you love that guy, but you gotta think about yourself. Your future." He reached for her hand and put his on top of it. "My advice is —just drop him."

She gently pulled her hand away. "No way. Uncle Charlie, I can't believe you're saying it. How many guys have you helped? The slouchers, the guys 'that just got mixed up with the wrong crowd.' Isn't that what you called them? And my guy, a hard-workin' reporter has been set *up*!"

"Whoa, whoa, Latanya . . . sweetie. Look, I, let's see. Did his family get him a defense attorney?"

"No . . . maybe. I don't think so."

"You haven't been in touch with them?"

How much can I reveal? Latanya puckered her lips and looked to the side.

"What, you don't get along with them?"

"No, I get along with them fine. They're good, honest folk, as you'd say," she replied, "They're not very well off. Defense attorney is beyond them, financially and practically."

"Well, country folk or not, the first thing you do is tell them to get a real defense attorney, and not the public defender—probably with a hundred-case load—he's got now."

Latanya nodded, staring into her cup of coffee. She waited ten seconds, until the silence was awkward, even between family members. "Uncle Charlie, could I borrow the computer and get it to the forensics people and back to the defender without anyone knowing?"

Uncle Charlie sat back up in his chair. He passed his hand over his beard and mouth. "Listen, Latanya, the law works in its own mysterious way. You can't push it—"

"But I gotta know, Uncle Charlie, I gotta know." She couldn't fight back the tears, flowing out as if the dam had been shattered. Her cheeks felt warm from them.

She felt Uncle Charlie's strong arms around her.

"Now, now, Tan. It'll be okay."

After another twenty seconds, she quieted down. She wiped her tear with a tissue.

He waited a while. "Okay, okay. Latanya, sometimes the family needs a break . . . I'll find out who the public defender is, and tell him to take this seriously. Spend half-a-day on his client, instead of half-an-hour."

Latanya took out a handkerchief and wiped her eyes. She smiled at Uncle Charlie.

"But will he be able to get the computer and have it tested?"

"Man, you're askin' for a lot there, Tan. An awful lot."

***

Jenkins' head bandage still bothered him. He could only see out of his right eye. As he exited, his left crutch rubber bottom hit the floor a quarter inch above the elevator cabin's floor. He stumbled awkwardly, just like any other patient who'd been on crutches for less than a week would, as he eyed the police officer near the desk through his un-patched eye.

He got out and stood for a few seconds, adjusting his weight, before hobbling towards the corridor leading to the room. As he entered the corridor, two officers at the end of the corridor looked up at him. It was funny, he thought to himself, that everyone hates a pedophile, but probably no one more than the policemen charged with protecting him.

With the crutches, head massively bandaged, he could have the run of the place. As he hobbled past the staff at the desk and continued, slowly, down the hall, he made note of the two cops' precise location.

He was more than halfway down the eighty-foot corridor before opening a door to his left and hopping in. One of the two post-op patients lifted his head and looked up at him.

"Good-day, fellas. Looks like Ah stepped into the wrong room, here," Jenkins said.

"Yeah? Well, that's not so bad . . . I think I'm in the wrong hospital," said the guy.

"And I swear I'm in Eastern Ohio. At least it smells like it in here whenever he farts," the other patient muttered.

Jenkins couldn't help chuckling while looked at the door between this room and the last one in this corridor, the one which held Robinson. It was bolted shut. No entry from here. He turned back to the two patients. "Well maybe iffen we're all so mixed up in here, we should ask the hospital to just move us to the Savannah Gentleman's Club down the road, and tell 'em to come and get us when they've figured it all out," Jenkins said, eliciting laughs from the first man and groans from the second.

As he stepped out, he pondered the facts. The room at the end of the hall was guarded by two cops 24/7. There was the side door to his room, but these guys would have to be neutralized and the door jimmied open, silently. *Not ideal. In fact it sucks, especially compared to the alternative.* According to the neurosurgeon he had chatted up, there was only a forty percent chance of recovery, which meant Robinson would probably never come out of his current vegetative state.

Of course, he wanted to be sure. With proper planning and a little negligence on the part of the policemen, he could organize something. But there was always a chance of getting caught, and more Law getting killed. Given the flack coming from the investigation, not such a good idea. Already, everyone involved in the 11/5 events was deep undercover . . . or deep in the ground. It would be better if the reporter just quietly stayed vegetative. Bueller's medic contact said a coma victim had a month, two on the outside to recover, to do so normally. After that, markedly greater chances of non-recovery. Time was on the side of 'The Project.'

He'd just have to make sure someone checked on Robinson's status every three days. He'd use his students. After a couple of months, it would be machine off and game over for the reporter. No fuss, no mess.

***

"So, what do you get when you cross an elephant and a rhino?" John Forrester asked everyone at the table.

They knew the answer, but saying so would've hurt Dad's feelings, or so they felt. So Latanya, like everyone else at the table, pretended to be absolutely dumbfounded.

"A Hell-if-I-know!"

Thanksgiving was a time of mixed feelings for Latanya. Good food and her dad's atrocious jokes, on the one hand; unabashed sibling rivalry on the other. The only positive on that front was that this year, her older brother Reginald had gone to his in-laws with the family. Uncle Charlie showed up with his wife Kokie instead.

"Honey, what is it about you Forrester's, that you can't keep your men . . . or women?" Aunt Kokie asked, looking at Michael.

"Now Mammie, you can't generalize like that, besides Latanya is one-half Tubman," uncle Charlie tossed in. "And anyway, when Mike gets married, he's gonna stand by his woman, ain'tcha Mike?"

Latanya's younger brother, a sinewy twenty-six-year-old with a nervous smile and a scar on his forehead, just laughed. "Oh yeah, Uncle Charlie, one bit . . . , uh, woman, for eternity." He took a swig from the beer mug. "One silky-skinned, tight-butted Rihanna fo' life," he clarified.

"You're dreamin' again, bro'" Latanya interjected, just as their mother brought in the stewed collard greens.

"Now, Tanya, this ain't no low-fat fine cuisine or that humma-whatever you fixin', but you want some home cookin' once in a while, dontcha, hon'?"

"Sure, Ma."

After supper, her mother pulled her aside in the side room.

"Tanya, sweetie, listen. I unnastan' you wantin' to stick wit da relationship, but if that guy is what the paper's sayin' he is, you just invitin' yo'self a boatload a' pain n' cryin'. Plus think o' yo' future."

"Ma, I told you, what the papers write just ain't true."

Momma nodded, leaned back and sighed, looking away.

"I gotta, I gotta go."

"Where to, honey? The turkey ain't even cold yet."

"I've got a lecture I gotta finish writing."

Latanya started to drive out of the city on Route 26 but turned around and pulled up in front of the hospital.

As the elevator opened up on the floor, she was struck by the quiet. She stepped out of the elevator, and a muscled female security staffer appeared from nowhere.

"Purpose of your visit?"

Of course she'd ask her. Latanya hadn't visited but twice since that time with Lieutenant Olson. "I'm here to see Paul Robinson."

Her response prompted the nurse at the desk to stand up. This nurse was different from the one she had seen two weeks ago, when she first stopped by.

The setting hadn't changed at all since her first visit. The female policeman, sitting in one of the chairs in front of Paul's room walked briskly to the front desk. The officer, a sturdy brunette with well-developed biceps, asked Latanya to step into one of the empty rooms. She shrugged; she would rather undergo the full body search, than hear that a crazed female had gotten in unchecked with a stiletto and killed Paul, a scenario she remembered seeing in at least three of her nightmares.

The two police at the door looked at her inquisitively. The woman cop got up.

As Latanya entered his room, she was struck by the quiet. The faint sound of the respirator rising and falling absorbed by the curtains. The screens were showing the line of his heartbeat; a steady, unwavering series of peaks.

She sat down on a chair next to the bed.

"Paul, I don't know if you can hear me, but it's Thanksgiving." She waited, hoping her words would cause some reaction. There was none.

"Well anyway, I can't stop thinking about you. I just know, in my heart, you're not all those things they say about you."

The only sound she got in response was the continued electric 'breathing' of the ventilator.

She took him by the hand.

"You rest here, while we figure things out."

She gently lifted his hand to her face and caressed her cheek with his palm.

"You hang in there. Don't go leaving on me, you hear?" She wiped a tear from her cheek.

On Tuesday, she entered the office of Ed Hanrahan, the 'more conscientious' public defender, according to Uncle Charlie. McKinley joined her in the public defender's quarters. The first thing Latanya saw, was about thirty manila folders stacked a foot high on his desk. Next to those, a Burger Devil bag, atop five legal pads, the top three askew. The smell of fast-food fries permeated the cramped space.

"We haven't received much evidence from the state . . . but what we've received is bad enough," Hanrahan started, before leaning back in the chair, and cradling the back of his head in his palms. "Your boyfriend was not personally about to rape the journalist, but he had one thousand and eighteen child porn images in his computer, and seven of those showed him explicitly having sex with them."

"It-it can't be," Latanya muttered feebly.

The public defender raised his eyebrows, shrugged his shoulders and turned to McKinley.

"Although that sounds shocking, it's not unusual for this profile," McKinley started, "but before we go there, let's say it is not entirely impossible that these may have been planted."

"I just don't have time for this hair-splitting," the public defender said, rubbing his balding crown.

"Look, you can't just let an innocent man rot in jail!" Latanya bawled out.

"Yeah? Well then hire a defense attorney." Hanrahan retorted, looking down at his cardboard coffee cup.

"The family doesn't have money, and--"

"Listen, Latanya, the first step is to eliminate the possibility of fraud or tampering," McKinley said, pushing his left index finger down with his right one. "If the forensic team has done their work well, then we'll be able to see just how many times he's accessed these files, how many times he's linked to others, et cetera. If the activity doesn't fit the profile, or we find some clue that it's been tampered with, then you can take the next step."

"And if it's typical?" Hanrahan asked.

"Then," McKinley turned to Latanya, "it'll be your call."

"I'll take my chances," she said, turning to Hanrahan.

"Okay, McKinley, what do you need? As a professor, you know that the evidence will only be turned over to me, and I won't be able to pass it further." Hanrahan glanced at Latanya. "The public—" he pointed to Latanya, "can't see it."

206

"Hmm. Maybe you could jot down a few key elements. There are only four," McKinley answered. "Latanya, at the risk of being cold-blooded, I'm going to be ultra-efficient. These elements will corroborate one another . . . or not. In the latter scenario, it will be difficult for the prosecutor to construct a case to go forward."

"And if the elements corroborate one another?" Hanrahan asked.

"Then—" McKinley turned to Latanya, "I can only suggest you walk away."

"Okay," Hanrahan responded, looking at his watch, "so what do you need?"

McKinley listed the search history needed on a legal pad.

Latanya sat looking at the far wall. *Why hadn't I ever accessed his e-mailbox? Other guys' girlfriends do that all the time.*

McKinley turned to Latanya. "Now, another element of this. I'm assuming you've seen lots of pictures of Paul. Smiling, not smiling, in various poses, from different years."

"Uh-hunh."

"Now, it's normal that you wouldn't remember most of them. But everyone might have three or four that they remember well. Stored in their subconscious, if you will." McKinley turned to Hanrahan. "Those seven pictures, Latanya has to see them."

"Impossible."

"Counselor, I'm not suggesting you scan them to her."

"No? Maybe I should cut them out for her."

"Counselor Hanrahan, you might just photograph them with your phone. Discretely. Show them to her, and then delete the photos. Nobody would be the wiser for it."

"I could lose my license."

"Edward . . . " Latanya broke in, "and so could a poor farming family in Loysburg. You can't believe what a hardscrabble life Paul's parents have been living since ninety-eight. And now, their only successful son's been accused of being a pedophile," she said, disgusted with the sound of the word rolling off her tongue. "If they had been Heinz's, they would have had ten attorneys by now. Paul pays for his father's dialysis. Mr. Hanrahan, come on . . . have a heart."

207

The public defender looked up at the ceiling and pursed his lips. He looked at McKinley and took a deep breath. "Okay, I'll do it, but my extra-curricular cooperation ends there. After that, he's another entry on my ever-growing list of ne'er-do-well clients."

"Thanks, Mr. Hanrahan," Latanya said.

"Ms. Forrester, note that I'm doing this only because your uncle got me out of a tight bind once. Don't abuse the favor."

"I won't. I promise."

"Now, if you'll pardon me, I have to consume this excuse of a lunch before my next court appearance."

Once outside, Latanya turned to McKinley. "And thank you, Professor McKinley. This is all so weird."

McKinley nodded. "Weird is what I teach to expect."

31

Olson and his team turned their attention to the Stoltzfuss farm the following week, when the snows had temporarily melted. Another reason was that the analysis had come back from the lab; one bullet had had the same striations on it – from a minor defect in the barrel of the gun – as a bullet found in the wall behind Robinson in Hollidayton. At this point, that put Rayman on the scene here. Unfortunately, forensic evidence had revealed no useable clues from the house.

Now, in late November, snow had blanketed the place again, and the investigation would slow. They had lifted, from the hardening ground, two items of note. Blood, not belonging to Stoltzfuss, found near where he was wounded.

Given the fact that Stoltzfuss' gun had recently been fired when they found it, Olson understood there was a good chance he had fired at the attacker. Since no slug from that gun had been found in the vicinity, it was safe to assume it had lodged in the victim. It had evidently not been a fatal wound.

The investigators had determined three other things about the scene. The first was that at least three different guns were fired in the vicinity of the fallen buck. The second was that the blood they found a few yards from Stoltzfuss had no DNA match on CODIS. And the third was that the spot where Stoltzfuss' blood lay was in the same trajectory as the fallen buck from the point where the thirty-odd-six shell was found.

But a hunting accident this close to habitation?

"Let's just say, hypothetically, that the frightened-but-guilty hunters approached him. He fired at one," Agnelli said, emulating a shot. "They fired back."

"Then what, they left without seeing if he was alive?" Olson asked.

"Yeah, that doesn't make much sense on the surface, unless they were from out-of-state" Agnelli answered.

"What about the lawyer and his car being here."

"The only thing we know so far is that the lawyer was killed by the same firearm as the captain at O'Reilly's house."

"Which would mean, according to our original findings, that Rayman was here and on Oregon Road."

Olson nodded. He turned to Finnegan. And you, what good news do you have?

"Rayman had no correspondence with the reporter and none with the journalist, before five weeks earlier."

"What about his computer?"

"Nothing there. But these pedophiles are no dummies. They use borrowed ISP addresses, download from the cloud. Any of 'em with a brain larger than a pea wouldn't use their workplace."

"What about the financials. Any clues there?" Olson asked.

"None. We looked at all his credit card transactions. Couldn't find anything on his cards linking to pedophilia, although one website was borderline, something about barely-legal shaved ones."

Olson winced. "How about cash withdrawals. Anything out-of-the-ordinary?"

"Nope. Perfect credit score," Finnegan added.

"Jeez. We just don't have a clear story here. Let's take this into high gear. Go to the shopping malls he hit in Penn. Get the scoop. Maybe he was approaching the youngest of the young."

Finnegan and Agnelli arched their eyebrows.

"Guys, I know this looks like a perv run, and we've been working from the premise that the two had set this up to make it look like it was linked to the voting machine thing. But ya know, it just could be that some vicious child molesters have killed two officers and disappeared into the night, leaving two more child molesters with smoking guns. And maybe one of 'em wasn't even a pedophile. Keep in mind, guys, we're not the Keystone Cops, we're the State Troopers."

Finnegan nodded.

"And Agnelli, look again at the kid's laptop. Finnegan, bring in Robinson's girl for questioning again when she's in town. She's gotta know something."

***

The next day, Latanya got a call from Hanrahan. At the assigned time in the afternoon, she went to his cramped office space. The side blinds were down, but now he pulled the front ones down as well.

"At the risk of people thinking we've got something going on between us, I err in the direction of caution."

"I understand," Latanya said, disgusted by the hint.

Hanrahan pulled out his Motorola Droid. He pushed some buttons, and two pictures came up on the screen. He looked at them and shook his head, then turned the Smartphone screen towards Latanya.

She winced. The top photo was of Paul in a missionary position with a girl of no more than ten years of age. Tears on her cheeks. The bottom photo showed him behind a girl of no more than eight, with his left hand touching her breast, while his other hand was groping another area.

Latanya jerked back. She clenched her teeth and shook her head. Then she lowered her head and looked again. He was kneeling. His features were sharper than the girl's. A camera focus issue.

"Can you stand anymore?" Hanrahan asked.

Latanya nodded, her insides screaming for her to walk away.

Two more appeared on the screen. The top one showed him from the back, while the girl, again no more than ten, was on a card table in front of him, her body angled to the left, and her legs on either side of him.

Latanya looked at the table legs. Then her eye wandered to the right. The synapses of her mind fired off. She had seen this photo before. *That was his sister's place . . . no - it was his uncle's cabin, wasn't it?!* The reddish-stained wood. The log cabin interior.

Hanrahan pushed a button and two more photos appeared.

"Wait! Go back."

She focused on the second photo. Paul was in profile. He was on his knees, bent over the girl's back, holding her by the stomach. He was looking towards the camera, smiling.

"That one, that one, it's a fake!"

"Shhh," hissed Hanrahan, looking towards the door. "How do you know?"

211

"Because, this was at his Uncle Ed's cabin last summer. I photographed these after we went swimming. We were both in the buff. And that bottom one-" Latanya pointed to with her finger, "His uncle had these bear skins all over the place. I was joshin' with him about - that being on top of the bear is better than being under the bear, and suddenly he held the bear skin up to his . . . um, midsection and said, 'On top, in all senses of the word.'"

Hanrahan pursed his lips. He nodded.

"I shot these two photos . . . who montaged them?"

"I don't know, Miss Forrester, but it seems you have the answer you need."

"I sure do!" Latanya answered, her pure joy mingling with fury. She got up. "Someone's playing us, goddamn it!"

"So, what are the next steps?"

She paced the floor twice, chewing her left thumbnail. She stopped and looked at Hanrahan. "No offense, Mr. Hanrahan, but I'm gonna get the best defense attorney in the state on this one."

"Miss Forrester, power to you. And as soon as you do, have him contact me, so I can hand over the case. In the meantime, here's the info that McKinley asked for," he said, handing her copies with figures and scribbled headings. "Now if you'll excuse me, I've got a baseball card thief to defend in ten minutes."

Latanya walked out to her car, still angry at the police that would have let him go to jail, would've let someone frame him, without lifting a finger. But even more than that, she was disappointed in herself. How could she have let others convince her of something that she knew wasn't true, of being within a hair's breadth of copping out.

She knew what she had to do next. She opened her phone and dialed a number she hadn't dialed in over six months.

"Hello?"

"It's me, Latanya."

"Oh, honey, we were just thinking about you," Paul's mother said, before pausing for a few seconds. "It must be hard." Latanya could feel the resignation in her voice. No longer could Latanya make out the jovial small-town grandma from a month earlier.

You don't know how hard. She hadn't told them the news yet. Although it might have made them happy, they would have been crushed by their inability to do anything for their son. She had heard from Paul that they were close to defaulting on a farm loan they'd taken out five years earlier, and too proud to take money from him.

"Listen, Mrs. Robinson, I'm calling with news."

"Tanya honey, that would be welcome just about now . . . we still can't figure out how Paul could have done what they say he has."

She was tempted to say: "That's what I'm calling about, Mrs. Robinson, he didn't." Instead, she bit her tongue. "We're working through it. Maybe they'll find he didn't . . . that it was a mix-up."

"What?"

"The defense attorney said the photos were odd . . . that they seemed doctored."

"Oh my, Latanya, that's nice of you to say," again pausing for several seconds. "But . . . are they sure?"

"Nobody is sure of anything, but the man seemed to have experience with this kind of thing," she lied.

"Well, when will they find out?"

"I don't know, they won't say. But I have one more thing I needed your okay on."

"Yes, child?"

"He's been assigned a public defender, in the event he wakes up."

"A what?"

"A defense attorney provided by the state . . . "

"Yes?"

"And this public defender's very busy and said he didn't have the time to build a convincing case, so he's recommended we hire a good defense attorney." Latanya bit her lip.

"Child, we, we just can't . . . "

"No, no, Mrs. Robinson, you wouldn't need to pay him. Paul has some money set aside. But since he's incapacitated, he can't commit the monies, unless the next-of-kin agree to assign someone the power-of-attorney over those monies."

"And you want us to assign it to you?"

213

"Yes. Now I wouldn't need them if Paul were not to wake up . . . but I'm so . . . I think he's just got to," she said, fighting back a tear.

"And darling . . . if he doesn't?"

Latanya steeled herself. She had pushed the thought aside, but had had to acknowledge that possibility. Once again, she felt like a lone tugboat captain facing a tsunami. "Then . . . I guess, I won't need anything."

"Oh, but honey, we'll keep praying for him, we'll keep sending prayers to the Good Lord for our little Paul. Just send whatever papers you need to. We'll sign 'em."

Latanya's insides twisted. She felt she was torturing Paul's parents. But she couldn't risk the information getting out until the defense lawyer was hired and had prepared a strategy. Suddenly, she felt she needed to give Paul's mom some ray of hope. "Claire, I, can't explain why, but I think . . . I think that Paul really needs for you to come here . . . I think that might, that might make a difference."

Her statement was met by silence.

"But John needs me to clean his apparatus, he needs to be fed."

"I understand, Claire, but just one day. A weekend maybe, when it's easier."

"I'll try, child. I'll try."

32

The next day, between lectures, Latanya stopped in McKinley's office, as agreed. He counseled her on what to emphasize to the new defense attorney.

She sensed she was walking with a lighter step. Now, at last, she could stop in and see him every day without shame or excuses. She'd already planned to camp out in his room over the weekend.

They had both seen the movie *Talk to Her*, and agreed it was one of Almodovar's best. The man who loved the woman, and talked to her, saved her life. As stupid as it might appear, Latanya would do the same for Paul.

On Saturday, Latanya arrived in the late morning to the hospital. Once again, she underwent the search, but this time she smiled, thinking how she could now boldly walk around this floor and confidently stand by her man.

"Good morning, officers!" She burst into a smile.

"Good morning, Latanya," the woman officer said. "You look happy today."

"Just feeling hopeful. I get a day like that once-in-a-while."

The woman officer nodded, while the man looked down to the floor and shook his head. "Well, if anyone can get it, it's you."

Latanya smiled, shrugged her shoulders and entered the room.

She sat by Paul's side and took his hand in hers. She held it like she had after the fourth time they had ever made love, when she was sure it wasn't just a fling. Now she bent her face towards the bed, turned his hand palm up, and cupped it around her cheek. It felt a little cold, but alive.

"Remember last summer, when we went to your Uncle's cabin?" she whispered, making sure the officers wouldn't hear through the crack in the door. "They faked the photos. The one where you doin' the bear." She sat up, half-expecting him open his eyes wide and say, "No way!" in his inimitable, ebullient manner.

His eyes remained closed.

She put his hand down. "But sweetie, rather than be sad, I'm gonna describe our favorite scenes from "Gone in Sixty Seconds." She did so for a quarter of an hour before feeling the urge to go to the bathroom.

Latanya, still daydreaming about their first encounter, didn't pay attention to the signs on the doors. She passed the first door on the left past the service elevator and the stairway, walked another twenty feet to the next door. She pushed.

It was locked, and a 'do not enter' sign was posted. She turned around, glimpsing a young blonde woman at the desk, dressed in a winter coat. The policeman at the desk was standing next to her.

"No change in his condition." the nurse said.

"Thanks," the blonde replied.

Sauntering back to the bathroom, Latanya stopped in her tracks. She did an about-face and walked briskly back towards the elevator bank.

Upon hearing the elevator doors open, Latanya started running, but she was still forty feet away.

"Wait!" she yelled, but the door shut before she was even thirty feet away. *She heard me.* Latanya stopped. *So young. A vigilante?*

Latanya turned to the nurse at the desk.

"Nurse, who was that?"

"Oh, she was from the Penn State paper. He had gone there, you know."

"Yes, I do. Thanks."

Latanya again walked back to the room. She made a mental note to investigate it next time she was on campus. If there were others who suspected Paul was innocent, that could only be good. But if it was some group of fanatical amazons hellbent on eviscerating him, that could spell trouble. And one thing she could be sure of, was finding both types at State.

The thoughts, triggered by the appearance of new players, kept nagging her. What was he really doing in Hollidayton? Why so far away? Had this been a *Twin Peaks* moment in his life, where he had shed his conventional suburban lifestyle for some sociopathic endeavor for a day, a week, a month? *Grand theft auto with a dead lawyer, to boot.*

216

The following day, Latanya arrived early. Her message to Claire Robinson had been brief. "Paul needs you, please come." She wanted to share the truth. It took Paul's mother a couple of days to arrange for some neighbors to sit for her husband, so she could visit only four days later.

When Mrs. Robinson arrived, that gray Sunday afternoon, she was carrying her Bible. After the awkward greetings, she confessed that she had half-lied to the neighbors; she told them she was going to see a relative in Harrisburg, as she had feared none of them would have lent her a hand to permit her to visit her 'pedophile' son.

Claire Robinson had said all this while looking down and thumbing through the century old Holy Book. The leather cover bore inch-long scratches, resembling scars. "Latanya, if Paul were back home, we'd all be goin' to church, so well, maybe we can do our own service here, absent Holy communion, o' course," she said after a half-minute of silence.

"Sure, Claire, sure."

What the heck. Gotta be thankful. And even if Paul went home only twice over the past year, well, it wouldn't hurt to make his mother feel good, too.

Claire took a hymnal out of her bag and opened it to a page with a dog-eared corner. She looked at Latanya as though she had been carrying rocks in her soul, and someone had just loaded on another ton.

"Latanya, would you mind singin' along?"

"No, not at all, but I'm not really musical, you know."

"That's fine, child. Like Father Salveson once said, in the Lord's eyes, you're an opera star if you sing these with heart."

They started with 'Baptize Us Anew'.

After singing a second one, Claire glanced at Latanya and smiled.

The two were then alone in their thoughts.

"Would you tell me again how you two met?" Claire suddenly asked.

Latanya recounted how Paul was visiting the Penn State journalism school, to give a visiting lecture on the keys of socially responsible reporting, which she decided to attend, on a lark. He had something about him, a generosity, a playfulness and a great wit. And he finished by saying, 'the two best things you can bring to the journalism world is faith in your fellow man, and doubt in your politicians.' "He was something special."

Before Latanya knew it, two-thirty was at hand. It would be almost a three-hour drive back to Loysburg, so Claire would have to return home now. Latanya accompanied her, as she walked slowly out of the hospital and to the nearby garage. Latanya waved goodbye. As Mrs. Robinson powered the 1997 Chevy out of the parking lot, she rolled down the window. "Take care of yourself, sweetie."

Latanya was glad Paul's mother had seen him. After all, she thought, as she surveyed the dark gray clouds gathering ominously to the West, this day, like any other this long into the coma, could be his last.

Defense attorney Fauntleroy was as self-centered as she had expected he would be. He'd win the case, not for Paul's sake, but for his own; he'd do anything to prove he was right. He was perfect for the job.

She told him that it would be hard on the parents to appear. They agreed that the sister would come out to Harrisburg. Linda Hoffman-Robinson would be the face of the family. She was tough-skinned and wouldn't succumb to the taunts everyone expected to be showered down on Paul until he was proven innocent.

Two days later, when Latanya returned to the ward, she was surprised to see a nurse in the room, as well as two doctors.

"When did that happen?" the first physician asked the nurse.

"I got the signal at two-ten," the nurse responded.

"Okay, so he's holding steady now for what, fifteen minutes?"

"Good, good. Keep monitoring."

The attending physician looked quizzically at Latanya. "Family?"

"Very much so, doctor."

The doctor looked at her down his nose. "Well, if the police cleared you, I won't ask. Anyway, good news. He's off the mechanical respirator. He started breathing on his own."

"He's coming back?!" Latanya asked, unable to conceal her excitement.

"No, we can't say that yet, but let's just say he cleared the first little hurdle of about ten. He could always slip up down the track." The doctor left and the nurse stayed for a few more minutes.

As soon as the nurse left, Latanya leaned towards Paul and took his hand. She held it tightly. "Oh, Paul, sweetie, please come back."

There was no reaction. She waited for twenty minutes, her hand molding with his while she recited every prayer she knew, twice. Not a movement.

Latanya took out a book. "I thought you'd need some poetry. Remember when you read Khalil Gibran to me on the beach?"

She paused, but the only sound that emanated was a low-level beep from the electronic monitor

"Okay," she said, opening to a post-it-noted page. She started on "Love one another, but make not a bond of love" when she heard the policemen get up.

Someone came into the room next door. The light came on.

"Jest lookin' for ma' friend," she heard outside.

"Let it rather be a moving sea between the shores of your souls," she kept reading.

The light went off.

"No, he ain't here."

"—Fill each other's cup, but drink not from one cup. Give one another of your bread, but eat not from the same loaf. Sing and dance together and be joyous, but let each one of you be alone, even as the strings of a lute are alone though they quiver with the same music--" she continued, before putting down the book.

The silence next door she found eery. She also needed to go to the bathroom. As she left the room, she saw a guy fifty feet down the hall leaning on one crutch, and hitting the bathroom door open with the other crutch. He looked inside.

Latanya heard the policeman at the desk get up.

219

"Need some help?" he asked.

"No, Ah'm alright, but thanks fer offerin'," the blonde man said with a Texas drawl.

Latanya froze. Even with only a one-third profile, she recognized him. She had seen him hobbling in front of the Eisenhower Auditorium. Some compassionate streak in her always forced her to notice the less fortunate and especially those in wheelchairs or on crutches.

Instinctively, Latanya turned around, so he couldn't see her face. She faced the policewoman to the right of the door. "Do you know what time it is? My watch stopped."

"Sure, it's thirty-four past three."

"Thanks."

But she had to continue to the bathroom and once there, laughed at herself aloud. *Like I'm the only one who could be living in Harrisburg and working in State . . . probably affiliated with one of the sports programs, so he has means.*

A few minutes later, still smiling, Latanya sat again at Paul's side.

The nurse re-entered. She adjusted some things, and took off the colostomy bag. Turning to Latanya, she reminded her that visiting hours were over.

As Latanya touched the door handle, the nurse added: "Honey, just don't get your hopes up too high. They sometimes wake up just to go straight back to sleep . . . and beyond."

Latanya froze. "I... I won't."

When she woke up the next morning, Latanya replayed the nightmare in her head. The guy with crutches was beating Paul with one of his crutches. Hitting him hard. Smiling while he beat him.

She poured herself a coffee, and then recalled the movies she'd seen, when the suspect in the hospital or under custody didn't make it. Witness elimination. *What if they poison the food, like in that 'Traffic' film? Pay someone to knock him off, like that episode in CSI-Miami?* If the photos were doctored, then it was a set-up, wouldn't the 'bad guys' have resources?

220

Her father and mother had sheltered her from the downside of the 'hood. She had gone to private schools, met good kids, learned mutual respect. Latanya suddenly realized she was woefully unprepared for these alternative scenarios. Her older brother Reginald, a patent attorney in the food business, would be even less helpful in this situation. But Michael, Michael—she'd never turned to him for anything useful, save the weed for her Master's graduation party three years back . . . *if one could call that useful. But oh, so many times she had bailed him out.* If it wasn't running to Uncle Charlie for help, then it was covering up for him to Mom or Dad. Or letting him crash out on her couch and giving him advice on cooling off the relationship, when he'd gotten mixed up with a ho' from the wrong crowd.

She looked at her watch and put down the phone. She scratched her head, then picked up the phone again.

"Yo' whatcha want?" the voice said groggily.

"Mike, it's me, Tanya."

"Sis, whachu callin' fo' at—" he yawned, "seven in da mornin'."

"Mike, I need some advice on something. Can we meet this morning?"

"Tan, you know how ta shake things up good."

"*Whazzup?!*" Latanya heard a female voice ask.

"Jez ma sistah callin', babe. You keep sleepin' now, suga'," she overheard. "Sis', I can meet you at eight-thirty."

They met at the Village Coffee shop. Latanya had dressed down for the meeting in beat-up jeans, sneakers and an old Steelers jacket.

Mike was dressed in the equivalent, although Latanya suspected he was doing all right with whatever he was doing. The Rolex watch he wore at the family dinner may have been fake, but the Armani calfskin jacket was not.

Latanya sat down in the chair with ripped patent leather back and chrome legs that had rust specks visible. She pondered how she was going to ask her brother for help.

Since the Thanksgiving dinner, Michael knew who Paul was in her life. But besides that emotional outpouring, what did Michael know of Paul? He had met him twice at the Phillies games, when Latanya had arranged a double date. Latanya suspected they would never be fast friends. *But life would tell.*

"Thanks for coming, bro'."

"Man, you must be needing something bad, calling me at seven o'clock. Good thing you didn't call me at six. That girl was steamin' . . ."

"Spare me the details."

"So lay it on, whatchu gettin' me up fo' so early? Da Law ain't even up yet."

"Cut the street jive. Unlike the trash you pick up at night, I know where you went to school."

Mike smiled broadly and shook his head, "Yes ma'am."

Latanya explained that Paul was in the hospital, and she felt he might be in danger.

Michael nodded. "Okay. So what do you want from me?"

"Mike, if his situation improves, I want you to steal him from the hospital and hide him somewhere safe."

"Yo, yo, yo! Are you crazy, or what?"

Latanya stood her ground, looking him straight in the eyes.

"You gotta be joking, there'll be cops all over the place.

"Three on his floor, two of them in front of his room."

"Yeah, and fifteen at the entrance—some undercover dudes. Shit, Tanya, those days are behind me. I've gone clean. Stealing, concealing him. Those are serious offences."

"Reginald will cover for you, if things go bad."

Michael burst out in a hearty laugh. "Tan, that's a good one. You're funnier than Chappelle. Only thing brother Reggie's gonna do for me is make sure the macaroni 'n cheese I get in the slammer wasn't made with Cheetoh's sauce instead of Cheez Whiz sauce. With help like that, I'm gonna be printing license plates for life."

The comment brought out a chuckle from Latanya, who needed one, after sleeping for a total of two hours the night before.

"Let the Law do their job, Sis."

"Mike, I saw a guy there in the hallway, this guy on crutches, and it seemed to me he was casing the floor."

"What movies you been watchin'? Stay away from that 'Bourne Identity' stuff," he said, shaking his head. "A man wit' crutches in da hospital. We done entered a new era."

Latanya couldn't help laughing again. She had a tough time accepting her younger brother's life choices. He had been given the same opportunities that she and Reginald had receivied, but he always seemed to get in trouble despite them. He didn't have to be an element of the streets, but he chose that, and the choice weighed heavily on her. But, she couldn't refrain from empathizing with him, unlike Reginald who seemed resigned— even enthusiastic at times—to see his brother punished. They also had mutual secrets, the most important being that she knew he peddled narcotics; mostly to high end, functioning drug addicts, such as the lobbyists, lawyers and anyone else serving the huge government apparatus of the Keystone State, and some even in it. But regardless the level of his clientele, she was pleased neither with his chosen profession, nor with the burden of having to keep it a secret from the rest of the extended family.

"Yeah . . . maybe, maybe it's just my suspicions acting up. I'm sorry, Bro', maybe I've been too, too, I dunno."

"Sis, yo' 'B' 'F's a no-good perv."

"Shush, Mike! Paul's been set up, and I think I know why."

"By Goldmember?"

"Mike, this is serious. You gotta swear to me you're not gonna tell anyone."

"What, that he peed in the fountain after eatin' asparagus?" he said, leaning back and laughing.

Latanya focused intensely on his left eye with her patented *this-is-serious-so-if-you-laugh-I'm-gonna-hurt-you* expression, the one she'd relied on dozens of times during his childhood to bring him into line.

He stopped laughing. "Sorry, Sis . . . I'm listening."

She explained the work that Paul had set off to do that day, and what Olson had told her last week. "The journalist was almost raped but then killed viciously, and the farmer who had a problem with the voting machine was killed in a hunting . . . *accident*," she made the quote sign with two fingers."

223

Michael was no longer smiling. He caressed his chin. "Dat's bad shit. Real bad shit."

"So??"

He looked up at the ceiling and scratched his beard. "But it's still too risky."

Latanya looked down at the table and nodded. "I understand. I should tell everything to the police and let them provide the protection then."

"Shit Tan, tell you the truth – the Law can't find their own butts with a map."

Latanya jerked her chair back. "That's a bit mean. What about Uncle Charlie?"

"Okay, he could, but with a *big* map. So many dudes I see that should be wearin' peels—"

"What's that?"

"Orange peels. The orange prison uniform. But they're not, these cold-blooded killers and armed robbers are just cruisin' the streets. " Michael scratched his head, slowly, deliberately. "Okay. But *if* I can hide him, who's gonna take care of him. You know they gonna be followin' you like a undersexed Ay-rab'd follow Beyoncé. You can't do that. And I ain't no nurse," he said smiling, "No mutha-f—."

"Mike! Cut it," she repeated, glaring. "I'll find someone; Auntie Mae knows some hospital staff . . . "

He turned serious again. "You think that part through. I'll scope out the hospital outside . . . but you gotta tell me about that floor where Boy Wonder's lying."

She started drawing the layout, right on the table. Where the doors were, the bathrooms, the desk, even the janitor's closet. Latanya would communicate the info on the stairways later in the day.

After, he explained what they might do.

"Thanks, Mike."

"Thank me when it's over. Oh, and Sis', when do you need this done?"

"When he wakes up. He needs the meds, the monitors, until he wakes up."

"When's that gonna be?"

224

"I dunno. It could be in a day, could be—"

"A day?! . . . You're crazy."

"As soon as he does," Latanya calmly continued, "the news'll hit the press and, well, I'm afraid, Mike, I'm afraid. Maybe it's nothing, but all these coincidences."

"Hmmm," he grunted, scratching his meticulously-shaved beard. "I'll need two days, if I can do it at all."

33

Jenkins was watching Flaxx News Monday night when the telephone rang.

"Arnie, what's your plan?" the voice asked.

"About what?"

"What news're you watching?"

"Flaxx."

"Then check out WQBS. The family's getting a new lawyer for Robinson."

"What?"

"Yeah, exactly. If Robinson wakes up, what makes you sure they won't be able to blow holes in the evidence?"

"But I've been checking every other day. His condition hasn't changed."

"And it best not. Arnold, I've got a specialist I can send you now. He works hospitals."

"I can handle this myself."

"Do you play chess, Jenkins."

"As a kid, a few times," he lied.

"Well, you can work with a knight, and if he's lost, well, you've still got the bishops, but you sacrifice the knight first. Besides, he works well in crowded conditions."

Jenkins was silent.

"Take him."

Jenkins hesitated. He was being pressed and didn't like that feeling. "Arright," he said, suppressing the irritation in his voice.

He hung up the phone and began pondering his alternatives. Only medical personnel would have access to Robinson. With guards by the door, they would have to be distracted somehow. Smothering with a pillow wasn't viable. Someone would have to inject something into the I.V. drips or directly into his arm; that would be the most straightforward, but with what? *Don't want someone telling me who I have to use. But who else?* Friday was Ely's turn to 'make a call on' her 'mentor.' Maybe she could change her profile to outraged woman avenger?

Jenkins picked up the phone and dialed. She answered quickly. "Hey, Ely, honey, how are you with syringes?" he asked.

"I think I do a pretty good job with yours, don't I?"

Jenkins couldn't help chuckling. "Touché, honey, touché . . . Ah have to agree, an excellent job." He stopped smiling and pursed his lips.

"So, will you be comin' to State tomorrow?"

"Yeah." Then he felt like he did when he had invited young Iraqi boys to walk ahead of him on the mined roads. "But seriously, Ely, you'd make a fine nurse—"

"Sorry, not in my plans," she interrupted. "I'm following in Michelle Beanman's footsteps."

"Hmm. Not comfortable with injections?"

"Un-unh. You should talk to Julie. She was studying to be a nurse, you know."

"No, I had no idea. Thanks for the tip. I'll see you tomorrow, sweetie, and my syringe's gonna be plenty ready for action in the ward."

"Naughty boy."

"Only with you, sweetie, only with you."

After he hung up, he felt a twinge of guilt. But the guilt swiftly exited. *Let 'em grow up fast, the world's no sitcom.*

He realized he had no choice. The boss had given the command, and he had no credible alternative; the 'clean-up' knight it would be. Jenkins couldn't help thinking how lucky that reporter had been. He winced. *At best, I'll be off the crutches at the end of January.*

He took out another of his disposable cell phones and dialed the hand-written number on the back of the "Humana" business card. "Ha-a, this is Arnie Jenkins. Is the organizer there?"

"Speaking."

"Ah'm gonna take up the boss' offer."

"Fine. This phone number?"

"Yep."

"I'll call you in ten minutes."

227

Jenkins hung up and walked to the end of the street on his crutches. He had no problem maneuvering with them by now.

He looked at his watch. Six minutes had passed, and it was really cold. He recalled one of his few cold moments in Iraq, when he had to assassinate a Kurd who was agitating too vigorously for his state's independence. He stayed overnight, ensconced between the rocks in the Northern sector. When the man showed up that morning to wash his hands and face at the well, he dropped him from three hundred thirty yards. The last thing the man saw was his own blood on his chest.

The phone rang and Jenkins picked it up.

"He'll be arriving on the 1:45 U.S. Airways flight from Dulles on Monday. He'll take a cab to the Hilton. Meet him at 2:30 in front of the bar on the 2nd Street side."

"Got it. Will he have the supplies needed?"

"Affirmative."

Jenkins hung up. He stayed in one spot without moving. He appreciated the cool crisp air right now; his leg didn't seem to itch as much.

***

Lieutenant Olson wolfed down the sausage and egg muffin. On top of the Tuesday government woes, the Patriot News had "Pedophile suspect wakes up from coma." He knew the city would bump up protection now, so it would just be a question of when the guy could speak. Olson called Finnegan.

Finnegan arrived a few seconds before Agnelli. "Well, Finnegan, I'm meeting the mayor in half-an-hour so we're gonna have to start bringing some conclusions to the table. Clear your schedules from this afternoon onwards, for the rest of the week."

"Will do, boss."

"Another thing, whaddya find out about the e-mails."

"The forensic guys couldn't get anything. The trash is emptied in cyberspace. It's not like a hard drive.

"You checked on his personal and office e-mail system?"

Finnegan nodded. "Yeah. Nothing."

"Did he tweet?"

"No. Not all journalists do. They've been burned with confidentiality breaches."

They finalized schedules and the two officers left.

He had one more thing to do. Olson picked up the interoffice phone and dialed. On the third ring, the police captain picked up. "Olson, you bastard you, right again on the Eagles."

"Just lucky, Jack, just lucky."

"Lucky, my ass. So, what can I do for you, besides financing your summer cottage, room-by-room?"

"Listen, Jerry, that reporter pedophile is waking up. You might want to station a few more people outside the hospital. You know how the crowds react when good things happen to bad people."

***

Jenkins had sketched out plans, reasonably accurate, of the last one required, the fourth floor of the hospital. He sat in the comfortable chair at the Toledo bar at the hotel. A guy wearing a blue blazer and topsider shoes approached, slowing down but not stopping.

"Jenkins?" the well-groomed, dark-haired man asked.

"The same," Jenkins said getting up.

"Don't get up. Fred's the name."

"A pleasure. I got the info you asked for."

"Let's order room service and dissect that in my room. Two-two-three."

In the room, Jenkins took out the plans he had carefully folded into his computer bag. He laid them out on the oversized desk.

"So where are the exits?" the well-dressed man asked.

Jenkins pointed them out on his map, and the executioner sketched the information in his pocket-sized notepad.

"So whatcha gonna do?" Jenkins asked.

"I dunno. I might dress up as a doc, walk into his room, inject the love potion and good-bye."

"Just like in the movies."

229

"But I might also pose as a relative, carrying a gift, and, when people around are relaxed, just inject some air into his veins." He brushed an almost invisible insect off his trousers. "There are hundreds of staff running around, visiting, studying, clocking in, clocking out. Security will have their heads up their asses as usual. Then again, if his condition changes, I might join the journalists. They'll be thick as thieves if anything changes. Nobody needs to be thinking who 'that doctor' was who just walked out before the patient died."

"How long before he dies?"

"You in a hurry?"

"Not really," Jenkins stretched his back.

"Two-three days?"

"That'll work for me, just fine," Jenkins replied, wondering where they found someone so expert in physiology.

34

When Latanya arrived at the hospital the next day, the sky was still gray, but lighter. In spots, the rays of the sun illuminated the rooftops and top floors of the few tall buildings of the city.

As she approached the room, it appeared to be a flurry of activity. One nurse came rushing out, carrying two vials. Another was holding the doorframe. One of the policemen had stood up and was observing the activity.

She felt her heartbeat quickening. This was the moment she'd been waiting for, but, she had to admit to herself, dreading as well. Latanya brought her middle finger to her lips, ready to chew the nail, but stopped herself. *Now calm down, girl.* Latanya positively sprinted the last thirty feet to the room.

Three nurses hovered above Paul.

"He definitely moved his toes," the tall one said, nodding.

"The monitor is showing a stronger heartbeat," said the one with salt and pepper hair, done up in a bun.

"What-is-it-what-is-it?!" Latanya, said, hurling her handbag into the chair in the corner.

"Good news," answered the second one, whom Latanya had befriended over the Saturday they had both 'kept watch' over Paul. "We have some signs of increased metabolic activity, which is good. We've got some brain activity, which is even better news."

Latanya smiled, then leaned over and, between all the wires and catheters, hugged him. "Paul, oh I knew you would, I knew it."

She stepped to the chair. "You may have been sleeping for a while, but I know that when you wake up, you'd want to catch up with...the world!" Latanya took out a New York Times, smiled to Paul and started reading.

After a few minutes, Latanya heard the sheets moving. She looked up, Paul's eyes were open. He was moving his hand back and forth along the sheet.

Latanya jumped up and held his hand.

"Paul, baby, it's me."

Paul uttered something that sounded like "sha-cha ga, sha-cha ga," closed his eyes, attempted to utter something else and fell back asleep.

Latanya saw the policewoman through the open door, but decided to say nothing. Then she pondered; what if something had to be done when he moved his hand or opened his eyes, something clinical?

She swung the door open and half-ran to the front desk.

The nurse looked up. "What did he do?"

"He said something," Latanya replied, "and opened his eyes."

"His eyes, really?"

"Yes!"

Within an hour, the first of three journalists arrived. They interviewed the doctor, then the nurse. Flashes from outside illuminated the room. Latanya heard the faint clicks of digital photo cameras. She moved her chair to the corner.

From what she could see, the police weren't checking the journalists thoroughly.

The last reporter repeated what others had said, almost verbatim: "Paul Robinson, suspected of pedophilia, and wounded during a shoot-out with police in Hollidayton that had left four people dead, has opened his eyes, moved his extremities, and according to nurses, has even attempted to utter some words."

The tall nurse, always sterner than the others, asked Latanya to leave the room.

Latanya didn't mind. She didn't feel she should be widely seen as his girlfriend just yet. She wasn't his wife in any official sense. But she still watched from outside the room. The male officer was now inside the room, while the female was also outside, keeping an eye on the journalists lined up outside.

One of the journalists, standing just inside the doorway, reached into his outer coat pocket. He was fingering something long and tubular.

232

Latanya stepped back into the room, slipping between the other journalists to get to this one. She grabbed the man's forearm.

Startled, he looked at her.

He continued slowly pulling out the tubular device.

She felt her neck muscles tense as bridge spans. She looked down to see a digital recorder.

Latanya stepped back, her the blood re-entering her head. "Excuse me, I thought—"

The journalist turned completely to face her. "And you are?"

"Just a friend," she said without thinking.

"A friend?" The journalist thrust the recorder three inches from her mouth. "And so, is there anything in his actions that tipped you off that he was a pedophile?"

This was not taking the path Latanya had expected or wanted. The policewoman had gone to the other side of the small crowd, so was no longer close enough to act. Latanya covered the top of the recorder with her hand, pulled herself in front of the journalist, and put on her iron lady face.

"I'm an undercover cop . . . You think a scum like that would have any friends?" she muttered through her clenched teeth.

The journalist smiled awkwardly, shook his head and turned back towards Paul.

As she passed by the front of the hospital that afternoon, Latanya was shocked to see forty or more picketers. The signs they carried read: "Protect Our Youth," and "Death to Perverts." She felt a tight knot in her stomach.

The next morning, Latanya woke up at 4:30. Another tough night, and now she was torn between sitting next to Paul and delivering the lecture. Latanya finally decided she had to deliver the lecture—her life couldn't come to a standstill. He would be okay. The nurse had told her that whatever she had seen in movies she could forget; the recovery would be slow. Paul wouldn't be speaking for days and possibly weeks, and if he did, it would be incoherently. As importantly, the steady stream of journalists would provide some protection. Any potential killer would be witnessed by them.

She would be back by two o'clock.

35

Paul's head hurt most of the time. He realized he must be sleeping twenty hours a day. The strangest thing for him though, was that he had no control over his bladder.

"Beep, beep, beep," he heard all the time. He had no choice but to be numb to it. He had also become accustomed to the police mumbling outside. Occasionally, he heard something about the Steelers. ' . . . great throw,' and 'missed tackle.' But after listening for ten minutes, he would find himself dozing off and waking up sometime later.

He had lost all track of time. Sometime yesterday, perhaps last week, he had gone through a harrowing experience. But the wounds he vaguely remembered receiving were all better. All healed, in fact, except his left side, which still ached a little. He couldn't reconcile the immediacy of the events and the fact that his injuries were all fine.

He recalled another tidbit from that day. A lawyer. Why would he think of a lawyer? Then he thought of a black BMW. *Not a car I'd normally drive.* It was all puzzling to him. When he tried to say something, all he could manage was a whisper, the words sounding funny. And he would sleep. He thought he had seen Latanya here, but wasn't sure. Nurses had been coming in and out all the time.

A doctor came in and did something to one of the bags suspended from the pole.

***

Jenkins waited in the car. Frederick was a pro. It was good that the boss man had kept in contact with people from Antoine's circle. Antoine had been very good. *How did he end up getting shot?* In any event, things were going swimmingly now. It was widely accepted that Robinson was guilty. Soon, he would be dead as well, and the world none the sadder for it.

He saw Fred coming out the side exit of the hospital, the employee exit, dressed in a doctor's blue scrubs. As he stopped, at the intersection, he looked in both directions. He continued walking briskly but effortlessly back to the car.

Jenkins appreciated the professionalism. Expertise at in the art of killing. Everyone has their niche.

As the man got in the car, Jenkins couldn't help probing. "How'd it go?"

"Fine." He pressed his lips together and thrust them forward. "It was so calm in the room that I decided to inject into the bag, not the line. Harder to trace. It'll take a little longer, but he wasn't going anywhere, according to the charts."

Jenkins felt a deep relief. He hit the steering wheel hard with the flat of his hand. "Yeah!"

The professional retained a Zen-like calm. "Is there a Korean eatery in town?" he asked as he put the empty pre-filled syringe back into a molded plastic case.

"Korean? . . . Oh, there's a Kim Suk Dik on every other street corner."

***

Paul awoke suddenly. Something had changed. He didn't hear the policemen talking in front of his door. Instead, he heard, in the distance, "Put down your gun." It was loud, but sounded like the policeman's voice. Suddenly, his door was flung open. Two black men came in, dressed in green intern's outfits, mouths covered by blue operating masks.

"How you feelin,' boy wonda'?" The taller one of the two asked.

Boy wonder?... Where's that from? Oh, Batman.

By the time he had figured that out, all his cannulae had been yanked out of his arms. A little pain. "Where's Batman?" he tried to ask.

"Wega Bammin'?" the black assistant with the beard responded. "Don't unnastand shit, ma man. Stop talkin'."

236

Suddenly, the two hospital assistants lifted him up. As they did, he heard, in the distance, "YESSIREE, I wanna know why dat slime pervert, who molested ma' li'l sistah, is bein' kept alive!"

Paul laughed, or tried to. That guy's mouth was definitely going to be washed out with soap. The interns carrying paid no mind to the swearing.

He felt himself being carried down the hallway. *To safety? Am I in danger?* He thought he remembered seeing two people outside his room. They were gone.

One intern held his legs, while the other had his hands under his armpits. The light was bright. He lifted his head up. As he looked to the end of the corridor, he glimpsed a black guy in a thick coat and gloves.

"You tell him, Martin! Why do the white boy dat rape yo' sista-who-be-nine, done get a soft bed . . . and three cops sittin' here. Riverside Mothafucka," Paul heard from another one he didn't see. He felt his body being juggled and twisted.

"Da injustice!" he heard from a third one in the distance, before he was turned to another angle and saw the top of a door frame pass in front of his eyes.

Just a small room with a high ceiling. More shaking. Why was the ceiling receding? Then the ceiling turned a quarter of the way around. Then another quarter turn. This was tiring. Another shake and his eyelids dropped down.

***

Latanya made her way past the immensely larger crowds than had been present a few days earlier, holding hostile signs. She walked through the front door of the hospital, she saw three policemen forcibly escorting a man and a woman out. "He deserves a stake through his heart!" the woman shouted.

Latanya began running to the elevator. A guard the size of a refrigerator stepped in her path and folded his arms. "Where you headed?"

"Fourth floor," she lied.

"To see?"

"My mother."

237

"Why the rush?"

Latanya turned halfway around and pointed to the crowd outside. "Them. I'm afraid for her safety."

"Don't worry. It's a lot of hot air here, everything's under control." He stepped aside. She smiled at him and proceeded to the elevator.

Latanya sensed her heart beating wildly. She took a deep breath. As the door opened on the third floor, she prepared to exit the elevator. A half-dozen policemen blocked the way. One stepped up to her.

"You can't get off here."

"Why not?"

"Can't tell you. Please continue going up or go back down."

She looked past him.

The officer turned to another policeman, who tapped his shoulder. Latanya used the distraction to bolt past him and to the counter. The nurse who took care of Paul looked at her, as she felt the arm of a policeman on her shoulder.

The nurse stepped from around the counter. "Officer, just a second. I know her, it's okay."

The officer loosened his grip.

"Latanya honey—I've got to tell you something. But best you sit down first," The nurse insisted.

Latanya stepped to a chair.

The nurse took her hand. She looked straight into Latanya's eyes. "Someone kidnapped Robinson."

"What?!!"

"About twenty minutes ago. He's gone."

Latanya grabbed her head in her hands, bent down, stood up, circled around and fell down.

She heard the nurse running around the desk and kneeling next to her. The nurse fanned her with the folder she was holding. *Excellent. Michael's come through!*

* * *

Paul woke up to see the black man in the front seat hold a gun to the driver's head, while the other one, from the back, put a

238

blindfold on the ambulance driver. They were in a warehouse or a garage.

"Listen dude. You take off dat' blindfold in less than five minutes, and ma boys outside gonna drill yo' head. We clear on dat?"

He saw the blindfolded man in the white coat nod.

"Good. And if you drive outta here in less than fifteen minutes, same thing. It's one-twenty five. Nod if you unnastand."

The man nodded.

Then, through the space between the seats, he saw the leader cut the plastic tie around the driver's hands.

The back door opened and the next thing Paul knew, he was looking at a gray sky above.

The snow was falling. But the fresh air was wonderful. They put a black jacket over him. It helped, but they didn't zip it up. All of a sudden, he felt very cold. Once again, they started carrying him by his arms, one on each side, but not his legs. *Why don't they zip up the jacket?* A minute later, he felt himself sitting in another car. He still felt cold. He was shivering. *Why am I not in the hospital?* He looked closely at the leader of the black men. Even with the surgical mask, the man looked familiar.

***

Olson was back in the office. The police commissioner would be down on Grover's neck for this one and just as hard on his own after that. He hadn't solved the case, and it now developed a corollary.

Seven cars next to and near the hospital had chased after five ambulances, a laundry service van and a taxi, but had found nothing. And those punks from the 'hood, even Harrisburg's relatively small one, had this uncanny ability to just disappear into the streets. The ones in the hospital had also been wearing gloves, so no useful fingerprints anywhere.

Olson waited for the police officers on that floor to come in. He looked at his watch. *Any minute.* He played this in his head thirty times, and what he really hoped for was that this reporter, who looked so normal on T.V. and whose stories had been so amusing, would now be fished out of the Susquehanna before the river froze over.

Captain Frank Grover started with descriptions of the black men. All in their twenties, stocky, wearing bulky coats, standard Timberlands. They could've been homeys from anywhere . . . black and dark blue jackets . . . Steelers, Flyers and Eagles caps. Two black caps with no markings. The toy 'weapons' had no prints on them.

They had questioned people in the hospital. The guys drew no alarm, as they had come in sets of two, or singly, some with flowers. Since it was about to snow, and it was cold, the gloves raised no alarms either.

Now someone they knew had Robinson, or knew where his body lay.

The female officer said that one nurse on the first floor had seen two interns carry a patient from the hospital to an ambulance, but she hadn't thought anything of it; special cases were shared with Holy Spirit Hospital all the time.

Captain Grover would be in hot water because his officers had abrogated protocol. One should have stayed with Robinson, no matter what. But in the heat of the moment, when five aggressive black youth show up, and one slips into the bathroom, hell, Olson wasn't sure he wouldn't have done the same.

But this could work out in his favor. When he collared Robinson, he would have some significant leverage over Grover. And he had no doubts he'd find Robinson, either in the flesh . . . or in the water.

After the brutal debriefing, Agnelli found him in the corridor. "We've found the ambulance. We've handed it over"

"Thanks." *A bit of hope.* They would work the ambulance over for evidence and find a hair, bullet casings with fingerprints, something. And then his team would circle in . . . mercilessly.

36

"Mike, I can't thank you enough," she uttered into the disposable cell phone.

"Yeah, Sis, this isn't something I'd do every week, not even every decade, for you. It cost me half a pound of the best weed I had . . . and I'll be calling in my biggest favors if I get caught."

"I'll pay you for it."

"Uh-hunch, Sis. On what, your wannabe prof's salary?" Michael asked. "Better you cook up 'deem collard greens wit' shrimp grits."

"Michael!"

He had stepped over the line. There was a moment of silence.

On her end, though, Latonya felt a little remorse from the outburst as well. After all, her brother had gone beyond bending the law. "Well . . . who can cook 'em up better den me, now dat gramma's gone, ah aks you?" she parodied back.

He laughed deep and long. "No one, Sis."

***

One day out and something gnawed at Olson. After another grilling from the state attorney, he went to the bathroom.

Suddenly, as he watched the water swirl, it hit him. *He'd need fluids.*

As Olson walked back to his office, Finnegan passed by.

"Finnegan."

"Yeah chief?"

"They're gonna need fluids. He's not eating solids yet."

"I got that, boss. We'll put every hospital, clinic and surgical supply store in the city on alert. Any new customers, or any disappearances from hospital wards, and we get a call."

"Good. Let's hope everyone's inventory system is up to the task."

241

Later in the day, he received a call from Latanya Forrester. She sounded distressed. He could understand that. The boyfriend was about to recover, when he's whisked away, perhaps to a murky death.

He could only console her with the usual. No suspects, no leads, but 'following up on it closely.'

***

Nurse Burns' comment that the three bags she got were 'not gonna hold him for even two days' rang in Latanya's ear. She found herself rubbing the small scar on her left index finger, the result of an altercation with her peeler-wielding older brother 23 years ago.

She wanted to end on a positive note, so she asked the nurse to put Michael back on the line. "How's he doing?"

"Well, the nurse changed the bedpan and your B-F mumbled something . . ."

"It won't hurt you to call him Paul, Mike."

"Yeah, yeah. I know he's heaven and earth for you. Anyway, since then he's been sleeping. The nurse said she'd be better with a monitor, but hooked up one of the last of the fluids anyway."

"Fluids. I know. I forgot about those. Damn it."

"Yeah, I picked up five bags and four bottles from a cabinet on the first floor, but we're gonna need more fast."

Latanya confirmed the what the suspensions were, quantities needed and rushed out the door.

She gunned the accelerator, passing all the cars in the left hand lane of the freeway taking her to Penn State.

***

Latanya turned off her phone. She took an off-ramp just outside of Harrisburg to Aunt Mae's house. Auntie was surprised to see her, but happy about the chance visit. Latanya used to stop by once a month when she was an undergraduate. Now almost a year had passed since her last visit. After drinking the last of the

242

coffee, Latanya said goodbye. She left her phone on the chair and pushed the chair in, to make sure Auntie didn't see it. She'd get it on the way back from State.

Latanya got on the off-ramp and was soon headed up to the University. Fifteen minutes later though, she pulled off the freeway and drove to Duncannon. She pulled into the parking lot of Ogilvy's surgical store and removed from her purse the first of ten prescriptions Nurse Burns had given her.

These were doctors the nurse had worked with. None of them had paid any special attention to her request. Lots of elderly patients needed this stuff. As the sweet old nurse had said, who would follow up on five bags of IV fluids, when twelve-year-olds were abusing cough syrup?

When she finished there, she drove further West. After two hours, she had thirty bags and enough needle sets to go with them, paying for all in cash each time. Finally, she could take off the thick-rimmed glasses and the constrictive headscarf. She also removed the fake mole on her left cheek. She smiled as she remembered one pharmacist's warning: "Better get that thing checked out. The sun's probably not as dangerous as where you come from, but it's still not safe up here."

37

Jenkins had pored over the papers for five days. Robinson hadn't been found, despite the fact that every police investigator in the state, it seemed, was on the case. It even took Jenkins two days to find out what had happened. Some black guys had created a diversion while someone stole Robinson from the hospital. Nobody had checked on him, except some ex-girlfriend. But his girls had never seen her there, and neither had he, so that must have been coincidental.

The whole situation sounded like it might even be a best-case scenario, where someone tortured and killed him. He hoped so.

Jenkins had gone over the science with Fred no less than ten times. The bean derivative needed half-an-hour to forty-five minutes to attain a lethal concentration in the body. Robinson had disappeared fifteen to twenty minutes after the injection. He'd probably make it through that attempt alive.

The transmissions they received from the police indicated the Law didn't have a clue. The five black guys who'd created a diversion had apparently evaded suspicion by coming in one at a time. They'd all been wearing gloves. The nanchuks the first policeman had been 'threatened with' were plastic. The 'pistol' the second one had seen was a toy gun. Jenkins knew that Pennsylvania's lack of laws against fake or toy firearms when confronting the law meant that the police couldn't have arrested them and the officers would have gone through hell if they had fired their weapons. All this was true, but that fact didn't make the next call easier.

"Yes, Jenkins," the voice at the other end answered.

"Ah don't know exactly what happened. I'm investigatin' it, but so far, it appears that a group of nigger activists took him in revenge for molestin' a relative."

"Jenkins, that's a fucking dream world scenario. You aren't so stupid as to hope for that are you?"

Jenkins paused a second. It's not often he was verbally slapped. This was no longer the brain trust giving orders. There was no more talk of 'reasonable' and the 'right thing to do.' Jenkins touched his disfigured nose. He felt something boil inside. "Hell NO, but what I am sayin' is that if I can't find him fo' five days and he shows up dead somewhere in an alley or a dog-fightin' ring, don't be too surprised."

"Jenkins, we're fine with a corpse gifted by someone else. Whether the Black Panthers arrange it, or you do, is jolly-fuck immaterial to 'the Program.' Someone just needs to fill another hole in the cemetery with the reporter's corpse!"

***

By mid-December, Paul was feeling more fit. He had gotten over the severe diarrhea, the headaches were less severe, and he could now think for two hours at a stretch before falling asleep. He wanted to tell his parents that he was alive, but he couldn't.

He didn't know where he was, he could only sense that this was in a lousy part of town. The roaches were all over the place, climbing in regularly through the open floorboard near the over-painted door. Sometimes he heard voices arguing in the apartment next door. Those people weren't nice. The nurse taking care of him was nice, though, and patient. *An angel of mercenaries. No, not sound right . . . mercators?*

Every waking moment, though, he was trying to piece together what had happened that evening in Hollidayton. Tan had brought him a whole boxful of Ken dolls and one Barbie, with which to reconstruct the scene. He played out scenario after scenario, but each time, the same question came up at the end. 'Who shot me?' He wasn't sure whether it was a thug who had come through the front door, or a naked man, or a policeman. Then he remembered something was going on to his left, after he had shot someone. Who was it? He didn't know. It was all a constantly deconstructing dream sequence.

***

245

Just after Christmas morning breakfast, Latanya slipped out. She went to the tenement apartment on 6th street the threesome had moved to.

"Merry Christmas, Paul," she said, once his eyes opened. "Hasha, slekable ona," she heard, nodding her head up and down. *Always be supportive.* It was hard on her, to see Paul, a robust entertainer with cutting wit reduced to a slobbering two-year-old, but she vowed to plug on.

The following Tuesday, while Latanya was re-connecting the fluid bag, she heard a feeble "tanks." Paul looked at her. She stared at him. "What did you say?"

"Tanks."

"Oh, Paul," she blurted out. "You're so welcome, honey." She covered him with kisses.

A tear trickled down his eye.

"I've missed you, Paul."

He smiled. "Har' t-talk."

"It's okay, sweetie, just say what you can."

Which was just another half-smile.

But the smile triggered something in her subconscious. A mild panic. She realized she hadn't called Olson in two weeks, when before that, she had been calling him every other day.

***

It was the day after Christmas, so it wasn't, Jenkins pondered, like they had ruined the holiday for the nigger. Really quite restrained of his man, he reflected, as he watched the massive Parkett wipe the brass knuckles on the white hand towel, already half crimson with blood. Parkett tucked the knuckles in his pocket and looked back.

Jenkins figured this guy would walk out of here on his own, but the bruised, red mass atop his neck would resemble an inverted fig for a week. "So, Jimmy, we're gonna check out that street. If I don't find him there, we're jes' gonna have to come back for you. But then, we ain't gonna be so patient," Jenkins said.

The bloodied face with one closed eye and lips swollen to the size of grapefruit segments nodded back, silently. Another drop of blood, mixed with saliva, fell from the lip onto the red-stained t-shirt.

"And remember, drop his photo for postbox 'three-one-eight' in the central post office."

They left, in possession of the information they needed.

Some guy named "Sam" had hired the crew, paying them in high-end weed. He'd also moved from his spot a few weeks back. At least he used the place for selling drugs a lot less often than he used to a year back. In fact, fig-face hadn't seen him there for six weeks. But now that they had a rough idea of where he'd moved, it didn't make sense to stake out 'Sam's corner.'

And stakeouts also didn't suit Jenkins anymore. Ever since his 'consulting' job in Iraq, he liked the feel of pursuit infinitely more.

***

"We've come up on five dead ends with this investigation, guys. Get TO WORK!" Olson sensed the fire in his cheeks. His investigators, sullen, got up from the chairs and exited his office.

None of the five black youths had been caught. One lead had heard they were from Philly. Drawings of their faces had been forwarded to the Philadelphia PD a week back, but the police there hadn't gotten back on them yet. This threatened to become a cold case, and Olson wasn't going to have one of those – not in this day and age.

Suddenly, a thought began to take shape. It involved Robinson's girlfriend. Why hadn't she called him to find out what had happened? She'd been calling him every day, or almost, until a couple of weeks ago. Now, silence. He thought back to her. She had allegedly fainted when she had heard he'd gone, but she hadn't called Olson but once since the disappearance. Maybe she knows something. The odd thought entered his mind - *does she have a relative, a close friend, who could have pulled it off?* He called Agnelli.

"We've gotta tap her phone."

"Whose?" Agnelli asked.

"Latanya Forrester's."

"Yeah, something fishy there, right. In an hour, we should have permission."

247

Olson nodded. "Very fishy.

"Hi, Lieutenant, it's me, Latanya."
Odd, how these mind waves travel, he thought to himself as he listened to her voice. But he wasn't going to recall the order about the phone.
"Yes?"
"Do you have any news?"
"So, you finally remembered?"
"Captain, I know it sounds odd . . . but I don't know if I should continue with this. I didn't tell you something."
Olson paused long, for effect. "What do you have?" he finally asked with the most jaded tone of voice he could muster.
"Robinson told me something, and at first it didn't make sense. But a week ago, I woke up and understood it."
"What was it?"
"He said, he said—" she hesitated, "he liked, he liked little girls."
"Is that exactly what he said?"
"'Like li'l girls'."
Olson permitted himself a long silence. "But you said you didn't believe it."
"I, I what can I say, Captain. I was hoping I'd misunderstood it. Paul was speaking in gibberish, half sentences. But after that, I just . . . " Latanya Forrester was sobbing. "I just couldn't pick up to call you. I just wanted to forget everything. Throw myself into my work. I didn't care what happened to him anymore. One way or another."
"I suppose I understand."
"So, if you don't mind, I'm not going to call you anymore."
"I understand."
"Good luck finding him. If I hear anything, I'll call again."
"Thanks. Appreciate it." Olson hung up the phone.
Finnegan came in. "Looks like she's driving to Penn."
"Fine. Call the medical center there as well . . . and their suppliers."

248

38

After New Year's, Latanya could sense Paul was improving. He kept muttering something about some machines could have been 'tamped with.'

Since it sounded closer to 'tampered,' Latanya decided to follow up on his mutterings. She called the State Department of Pennsylvania and inquired as to whether any provisional testing was done on the voting machines. Nobody seemed to know, but after receiving the third 'please hold-the-line while I transfer you,' a Ms. Perrino answered. She seemed enthusiastic and engaged, for a government official; she set up a meeting for the following Wednesday.

Latanya had not been very careful before Paul's 'kidnapping'. But now Michael's fate also depended on her being extremely cautious. And although she'd been calling Olson regularly for a few weeks, she had stopped calling him. *Maybe too abruptly.* The tone in his voice on the phone had been unpleasant. Sooner or later, they would have turned to her anyway, but now she couldn't help thinking that her 'gig was up'. The excuse she had given him about Paul's 'confession on little girls' was rubbish. She earnestly couldn't be sure that it would have fooled many people who knew her. But the delays had bought her time. Time that she could now use to pursue what Paul had been pursuing.

After her Wednesday morning lecture, Latanya left her iPhone in her office, and took her disposable clamshell instead. Furthermore, a colleague was out of town until the weekend and had asked her to mind her cat. Latanya took the liberty of using the assistant prof's car, just in case her own would be tracked.

She also took a few business cards she had printed out for this occasion. Her name was Florence Smith, and she was a polling specialist of a small IT Consultancy sub-division of David Greene's company. He had agreed to put his phone number as the business phone in the card, to lend credence to the facade. Her mobile number was the disposable clamshell. They had agreed that for the next few weeks, Greene would answer his phone with the name of his company, to cover. 'Florence' would be reachable only on her mobile number. If someone was very persistent, then Latanya would come to Greene's office and call from his phone.

Maria Perrino was petite, with dark hair and neatly-cut bangs. She boasted a warm smile.

Latanya handed her business card to Ms. Perrino. "Florence Smith, from 4Q Consulting. Elections and polling sub-division. You can call me Flo'."

Perrino nodded, still smiling. "How may I help you today, Mizz Smith?" she asked.

"I understand you were responsible for the 'parallel testing' of the voting machines," Latanya started.

"Yes, that's correct."

"Before I start with my specific query, could you tell me exactly what parallel testing is?" Latanya asked, smiling widely in return.

"Sure. The purpose of the parallel testing is twofold—" Perrino started, brushing a pine needle from her jacket. "The first is to assure that the voting machines are functioning properly, and the second, that they have not been tampered with. This is accomplished, by running a fictitious vote on the machines."

"And where are these machines taken from."

"Randomly, from precincts all over the state."

"Okay, and who does the parallel testing?"

"Volunteers . . . and if there aren't enough of those, then my staff."

"Hmm."

"So they will be emulating a day in a busy voting precinct, for four machines."

To Latanya, she sounded a little like a drill sergeant. Probably a good thing.

"Every step taken by the individual voting on the machine will be exactly the same steps a voter will take."

"Is there any cross-verification?"

"Sure. First off, when the ballots are read off, one of the four persons at the station tallies them manually. Secondly, all the activity will be captured on videodisc and reviewed if there is a discrepancy."

"Corroboration sounds pretty comprehensive." Latanya nodded slowly. "So, how many votes did they have to enter, a hundred or so?"

Perrino smiled. "Try four hundred and fifty."

"Wow. And were there any irregularities at any of the stations?"

"No."

"None? Not even a couple of votes off?"

Perrino leaned back and squinted. "You're writing a report on this, right?"

Latanya shook her head. "Nope. Not this specifically. Perhaps we'll profile the volunteers, as this represents heightened civic activity. But I'm not really concerned with small technicalities," she lied.

Perrino smiled warily. "Okay . . . now, each page at the stations represents a cast vote, and the vote must be read out, in the order appearing on the pages; first the U.S. senator, then the U.S. representative, followed by the state senator for the district, and so on,"

"So they can't reverse the order."

"Not that it would change things, but it would just make it that much harder to re-check it. Remember, Flo, we are talking about watching and pausing through eight hours of video. Two days minimum. If, on top of that, they're changing the order, then—"

Latanya waved her hand. "I could imagine."

"Remember, the second person would be accumulating the votes on paper, with a pencil or pen, so a cross-check is in place. The third person would be voting on the machine."

"Yeah. Doing this for eight hours could feel like forever."

"Exactly, so we have a fourth person at each station, who would rotate back in and take the vote reader's place every fifteen, twenty minutes. The new fourth could get coffee, go to the bathroom, stretch his or her legs."

Latanya looked at her watch. She would be late for the appointment with one of her students. "Well, thanks, Mizz Perrino, for the explanation. I'll tell my grad st—"

Latanya caught her error mid-sentence. "Assistants, about this, process today's info and make another appointment with you." She was about to go, when an idea crossed his mind. "Would it be possible for me to look at the videodiscs?"

Perrino frowned. She started shaking her head. "They're in storage."

Latanya creased her forehead.

"But I suppose we can take one out, Mizz Smith. A double-check."

"Honey, what about anything beforehand. Anything that day?"

"Tunnel. We in tunnel."

"Who's we?"

"Don't remem . . . ber. Com-com-p guy . . . hit other guy," Paul struggled, waving his hands. "Br- Blonde."

"How did he hit him, with the car?"

Paul shook his head violently, like a three-year-old denying he had pulled a cat's tail. "I driving. He was on floor." Paul stopped speaking. He was acquiring that glassy-eyed stare again, the one that told Latanya he was zoning out.

Ten minutes later, Latanya again continued her gentle prodding for information. After fifteen seconds, Paul realized something.

He smiled. "I remember. Cop-ter guy in tunnel hold- holding bat."

Latanya scratched her head. They were in a tunnel, a dark place. "A bat, as in dark, furry and flying around?" Latanya asked, swinging her arms in a flying gesture, testing his sense of humor.

He stared at her and wrinkled his forehead. He burst into laughter, then shook his head.

"Baseb-b-bat."

"Okay. Baseball bat."

Paul nodded. "He hit guy had gun."

"And then?"

"Car in front. Look at scra-we-llen."

253

"Scra?-"

He gestured a rectangular shape. "Scr- scr-"

"Screen!"

Paul nodded vigorously. "Backing up . . . red lights. Bump."

"Bump? Did you bump him?"

Paul stared hard ahead. He smiled like a child opening an unexpected but highly desired present. "Wipe out. Took gun. Slakrkin, ona," he uttered, before staring into space for a minute, then closing his eyes and falling asleep.

After two days though, Latanya had some major pieces of the personality in place and a solid answer. This guy had a broken hip because Paul had hit him with the car.

Latanya was pleased with how open Perrino was proving to be about the parallel testing process.

There were two teachers, who'd come every election, two male retirees who loved to golf and to round it out, two librarians. This was supplemented by seven college students and a lacrosse coach. Perrino said it was pretty rare to get college students.

"Why?" Latanya probed.

"Well, it's not that close to any major university, and parallel testing is not like working at the polls or, or getting out the vote. They are putting in an eight-hour day just to check some machines. They aren't campaigning, phone-calling to promote a winning candidate, you know, making history."

In response to the question of any other details, Perrino responded that one was about to graduate, and with no job offers. No other particulars, except that the students looked up to the lacrosse coach, which made sense, as it appeared he was the organizer.

"Great, so that settles the issue of profiles. Let's look at the process."

"Fine. So Flo', as I'd mentioned last time, we started with ballots. Here they are," Perrino said as she slid four piles to Latanya.

"Mind if I look through them?"

"Please, go ahead – I'll be sitting over at the desk by the door if you have any questions."

"Sure. Thanks."

Latanya went through some of the pages in the first pile. Squares next to candidates were marked off with checkmarks. Same with the referendums. Latanya went through about ten pages and saw that they had variations in votes, as would be normal.

Just to make sure they were all done the same, she glanced at a few pages on the next pile. Then the third pile. But as she glanced at the second page on the third pile, she noticed the super white discoloration associated with correction fluid. She ran her finger over it. Rough. It had been 'corrected.' An 'x' was placed in the Dooley box for senator. She continued looking through that pile until she found another page with a correction. There, once again, the box for candidate Duquesne was whited out, while the one for Dooley had the 'x' in it. Below, on this page, the referendum was also whited out.

She pondered the meaning. It didn't make sense really, because if this was just a test, it didn't matter how the votes added up. There would have been no 'total' to shoot for.

"Maria?" Latanya called out.

Perrino put down a document and came over. "Yes, Flo'?"

"I just came across this pile. It had like four pages, in the first twenty, with votes 'whited-out'. Would you know why that would be?"

"No. That's weird . . . let me see who filled in these ballots." Perrino shuffled through the pages in the pile. She pulled one out about forty pages down. "It's Brad. Just a second."

A few minutes later, Perrino re-emerged, followed by a lanky, disheveled young man, in tow. "Bradley, this is Florence Smith, from a consulting company here and I was just showing her the ballots. She noticed that there were 'corrected' ballots in this one pile. I recognized your handwriting on a later page, and these were all completed with your ink pen.

"Hmm . . . I don't know. I corrected one—because the write-in candidate's name was . . . inappropriate."

"Name?"

Brad smiled, "Ms. Perrino, it was boring. We needed a laugh."

255

"Okay, we'll discuss that later. So why so many? Four in the first twenty."

Brad scratched his head. "No way. It wasn't me. Why would I do that? Remember how we wanted to just finish A-S-A-P and get out of here?"

"Yes, that's true, it had been a long week. You can go. Thanks."

Perrino looked at Latanya, deep in thought.

"What about another scenario?" Latanya piped up. "Let's say two of your parallel testers are left or right extremists."

"Most probably the right, in these parts."

"Okay. Right, let's say. Even more logical. Leery of government. Might even think the election is rigged. His 'Conspiracy-Think' drives him or her to believe the ballots were prepared and matched up with the machine."

"I think I see where this is going."

"That's right – so he and his buddy change the votes on the ballots to throw the 'system' off."

Perrino nodded effusively. "Sure. I have to say, the golfers certainly were right of center—judging by how they were quoting Lush Trimbaw, trashing the president's policies."

"We could probably see if they were changing something on the ballot on the videodiscs."

"You're right, and you know, one of the golfers was the only volunteer to call back asking about discrepancies."

Latanya nodded. "Could we look at the videodiscs?"

"Not now, they're in storage."

"All right. I'll come back later." Latanya knew that by asking more about the whiting out, she might be taking the cover off this supposedly sociological study. "On a completely different note, did the students in the videodiscs strike you as being from one socioeconomic group?"

"Yep. The white tennis and polo set. You know, like they were going to joke about it over gin-and-tonics the next day."

Latanya took some notes, thanked Perrino and left.

Latanya promptly started researching and the 'blonde man' in question wasn't faculty. This left the sports clubs, sports agents, the gun club and two or three political organizations. Through asking her circles of colleagues and acquaintances, and by engaging in a process of elimination and deduction based on where he was hobbling, she had limited it to the fencing club, the gun club, the gay alliance, the Young Republicans and the Corn Cob Caucus organizations. He wasn't an agent. She felt she couldn't get any closer to knowing without possibly revealing her quest to the man.

She decided she would launch a social anthropology study on the politics of disabilities. She created a questionnaire. She carefully asked the organizations to profile, at the start, whether anyone in the organization, or among its advisors, was disabled, even temporarily so. She drew two boxes. Next to the top one she wrote 'short-term disabled.' Next to the bottom one, she wrote 'long-term (two years +).' It took her two hours to draft it on Excel. Rubenstein arrived at four, and she went over the questionnaire with her.

"And this feeds into the study on politics in rural and exurb communities?" the assistant asked.

"Indirectly, yes. We need to include this component in it, to eliminate alternative parameters. We need to eliminate physical handicaps as a reason for not voting. I'm also trialing the questionnaire format, so if you get questions on the survey questions, I'll need that feedback. We'll need to perfect it before we roll it out."

"Sure, sure. I get it. I'll start tomorrow," Rubenstein responded.

39

The following day, as Latanya walked to her car, she had the distinct feeling she was being watched. As she drove along the street towards Paul, she noticed a blue car following. She turned onto Maclay Street, and it followed behind her. There was *nothing* on that street. Nobody would take it. Latanya thought fast, and continued heading towards Front Street. She had recently concluded a project studying social indicators of ROTC enrollment at the Navy and Marine Corps Reserve Center. It would not be outside the realm of reason that she would be taking this small road to get there, It was frustrating to her, though, now that she would have to spend the next hour at the Center, pretending to follow-up on the project that had been concluded and evaluated ad nauseum five months ago.

***

"I hate following these grad students around. They have a schedule as predictable as a Philly alley cat's," Finnegan commented, between puffs of smoke.

"And nowadays they're about as useful as one too. Whatever happened to Math and Science . . . Engineering?" Olson took a deep breath. "Keep tracking her through the weekend. If nothing comes up, then we'll let it rest. No sense wasting more taxpayer's money. What about Michael Forrester?"

"He hasn't been at his spot. At all. Everyone thinks he's changed professions—"

"You have gotta be kiddin' me."

"He got roughed up badly during some fight six months back. He had told everyone he's gettin' 'too old for this stuff.'"

"Just pull him in next time he's on his spot," Olson concluded. "He'll be there sooner or later . . . doesn't look like his parent's trust would kick in yet – they're both alive."

Latanya didn't visit Paul the next day as well, out of fear that they'd track him down, not sure who the 'they' were. But on Thursday, she went again. Paul's morning 'gibberish-offload', as she had termed it, was interrupted three times with a clear enunciation of the word 'hummus.'

But Latanya's gentle querying had revealed some more information. It was clear that he had been at the programmer's place and there saw a dead body. He had fought with someone, then left with the programmer in the car. They had been followed and 'bumped' by a helicopter, but this was still sketchy. The car went into a tunnel, and then the programmer knocked one of the attackers with a bat, after which Paul had hit or run over the attacker with the car.

Subsequently, they had gone to another house, where a man was shot, then two more men were shooting, a driver was knocked unconscious, again by the programmer, and then things got fuzzy.

But as frustrating as the information gaps were, she noticed he occasionally enunciated whole sentences.

She wanted to take this information and leave it to Uncle Charlie, but she needed names, or at least precise descriptions. So far, she had one only of the assailant in the programmer's apartment, and the programmer himself. The former didn't matter, since—she researched—the assailant had been found a day later in an alley a few blocks from the scene of the crime. The junkie had robbed the programmer as well as killing the social worker. He must have been very high, because he had actually tried to hide the gun by stuffing it between mattresses.

Relying on Michael's drawing skills, a holdover from his high school days as an occasional cartoonist for the school paper, Latanya asked the both of them to come up with pictures of the persons remembered.

The fourth time, she confirmed with Paul that the drawings were accurate for the programmer.

"What about the guy you hit with the car?"

259

"Blonde-hair, big…spel forces ga," Paul replied.

Michael started drawing a sketch.

"What about his nose?"

"Straight, thin regu-, regu-lar."

"What about shape of face?"   "Squa-, Squa," he scratched out. Failing to complete those words, he said: "Schwarzenegg."

Michael erased what he had drawn, then drew in a square jaw.

"Eyebrows thick or thin?"

"Thin. . . Regu…nose crushed by bat. Thin. . .mints, thin mints."

"Sis', your man's zoning again."

And as they had done a dozen times that week, they quit working. Latanya sensed Paul was ascending to a cloud again, someplace no one else could glide to. She had to calm herself down. *He's recovering, he is re-co-ver-ing.*

Upon returning the next day, Latanya had a composite drawing. As Paul nodded off to sleep, she scrutinized the photo. *Was this the guy on the crutches?* Latanya recalled the first time she had seen him, getting off the elevator in the hospital. He still had a bandage on his head, and a splint on his nose. Later, she asked Paul again where the guy had taken the bumper and he had responded 'in Leg', 'Up Leg'.

Paul woke up and Latanya continued. He pointed, with certainty, at his upper leg. With a glint in his eye, Paul added: "Took gun."

"You took his gun?"

Paul looked at the wall for a few seconds, and nodded.

"Tired," he said, a few seconds before closing his eyes and falling sound asleep.

Latanya looked at his face, more furrowed than it had been a month earlier. "Michael, does it sound to you like the guy—this square-jawed guy, was hit by the car hard enough to lose consciousness… maybe his weapon?"

"Yeah, could be."

"Then that could be the guy I saw – the one on crutches."

"Sis, now don't be jumping to conclusions."

"I won't, I'll bring a photo."

The following day, Latanya arrived with a photo she had downloaded from the Corn Cob Caucus site.

Paul was restive when she arrived. He moved all around the bed. "Want up!" he said. He pulled out his feeding tubes. "Want eat ha-, ha-, burger."

"Honey, honey, you will, you will. Just don't do anything crazy."

"Ain't gonna keep it down, no sirree," said Nurse Burns, shaking her head.

"But will it harm him?" Latanya asked.

"No, honey, it won't."

Paul's eyes burned angrily at Nurse Burns.

"Shit, shit, shit." "Okay, honey, I'll be right back with a burger."

His eyes lit up as he broke out in a childish grin. "N fries!"

Latanya couldn't help giggling.

"Greasy ones?"

"Not fu-n, fu-nny!" Paul yelled.

"He's havin' one o' his 'I'm pissed with the world' days, honey, don't pay him no mind," Nurse Burns interjected.

Latanya nodded, then turned back to Paul. "No, it's not funny at all." She faked a frown. "But before I go, can you tell me if this is the guy you hit with the car?" she said, pulling the photo of Jenkins out of her bag.

Paul looked at it. His neck stiffened. He started hitting his hands against the bed frame. "Yes, *yes*."

"Okay, honey. I got that."

Paul started rocking back and forth.

"How about a milkshake?!" Latanya asked.

Paul stopped rocking and looked up at ceiling. He looked down. "Oh yeah," he nodded vigorously, "Chocnilla!"

Latanya bent over and bit her tongue to keep from laughing out loud. She turned to the door. "Be right back," she called out.

She walked out the inside door, down the dank hallway to the outer door, passing two other doors in the process. Behind one door, she heard loud yelling. "He almost O dee'd. What dose you give him?!"

"Da usual shit, mutha fucka."

261

Latanya stepped out quickly. This definitely wasn't the neighborhood she'd grown up in. She ducked down the alley, hoping that she had done so fast enough to avoid being spotted by someone, and walked to the car.

Latanya wondered, as she walked down the street, why she needed to continue hiding Paul. She had a proof that the photos were doctored, she had identified the culprit in the killings, and besides, what she was doing was illegal and quite dangerous to her brother. Besides, nothing had actually happened to Paul in the hospital.

She decided she would not just tell the police where he was, but to actually bring him into the station, to Olson. She'd decided to do it in two days, to give time for the nurse and Michael to prepare him.

But something tugged at her. Just slightly. Nobody had said anything about his desktop computer. The police had stopped with the laptop . . .

When she returned with the bag containing a cheeseburger, fries and two shakes, she heard Paul yelling "Burger, burger. Lat . . . anya," as she put the key in the door,

She opened it and saw him sitting up, waving his arms back and forth, as the nurse held a butterfly needle attached to the feeding tubes up, attempting to insert it.

"Paul, stop!"

Paul stopped waving. He looked at her with a puzzled expression and stared at the bag. His gaze softened. He smiled.

"Mrs. Burns, sorry for this."

"He may want that burger, honey, but like I said, he won't keep it down."

Latanya paused, shrugged her shoulders, and then turned to Paul. "Paul honey, I can give you this burger, but only if you promise to let Nurse Burns put in the feeding tube needles."

Paul squinted. He gradually relaxed the squint, and sitting back up in the bed, nodded. Nurse Burns proceeded to put the butterfly needle into the vein, while Latanya took out and unwrapped the burger.

Paul's eyes lit up. He reached for the sandwich and jammed it into his upper lip, then lowered it and bit off a quarter of it.

"Don't choke on it, honey." She helped him consume it, as best she could. At the end of it all, the bed sheet was covered with ketchup splotches and chocolate-vanilla shake, highlighted with bits of pickle and a triangle of American cheese.

"Mmm-Mmm," Paul purred triumphantly.

"Just one more thing, sweetie. Do you remember your desktop computer password?" Three times he had given her a different one and three times it had been wrong.

Is he hiding it from me?

40

Once again, Latanya borrowed the grad student's car and returned to the DoS. She viewed the first of the 12 discs, three per table, provided by Perrino. Perrino would join her after a meeting.

Latanya started watching the first parallel test table. She noticed nothing odd, aside from one of two elderly ladies, who appeared to have trouble voting. She kept criticizing the machine. The second videodisc had more of the same.

She switched to watching the videos with the two golfers and two librarians. Four retirees, but these appeared to have the system down. One of them, before the second round of voting started, yelled out the following: "Did you like that about the President flying to a conference in Macau last week. He didn't visit any of the casinos. You know why? . . . When you're $10 trillion in debt, they don't let you play." The others laughed. Latanya looked intensely at the tape. The men and the librarian appeared relaxed.

After Latanya had been watching the tape for about thirty minutes, Perrino showed up.

"Hi Florence, how are you?"

"Good, thanks. Started with the two schoolteachers, now watching the tape with the golfers. Nothing out-of-the-ordinary."

"Yes, there shouldn't be. I meant to tell you in the note, the ballots with the corrections were from Table 3."

"Table 3?"

"Yes, the three students and the lacrosse coach."

"Thanks." Latanya quickly replaced the DVD's. At table three, she saw two young women, obviously coeds and a male student she estimated to be an upperclassman. One of the coeds was voting on the machine, while the other one wrote down what the male student said. The angle of the camera precluded seeing the fourth person.

The team voted about twenty times. Nothing out-of-the-ordinary. Suddenly, the woman voting backed up and moved to the left, blocking the camera lens for about two minutes. Then she turned sideways and faced to her right. In the video Perrino showed up, asking what the problem might be. The girl moved farther to her left, and a new but familiar face appeared.

Latanya suddenly felt cold sweat on her back.

He now boasted a mustache, but it was unmistakably him, square jaw, blonde hair and all. Latanya wound back the videodisc to one minute before the blockage of the camera. She reviewed everything. *Nothing strange.* The stack of voting cards was untouched, the writing with the pencil the same. Then she noticed the pile of ballots—something was different before and after. Again she wound back the videodisc. There she saw it. The pile was lower, by about half-an-inch, before the coed had blocked the camera.

A thousand thoughts raged through Latanya's mind. Key among the, was that somehow, the voting machines were the link to the gunfights leading to his current state. Latanya then stepped back. *How did they manipulate them, exactly?*

Perrino walked through the door at that moment.

"Hi Miz Perrino, good to see you. I just saw something odd."

"Hi Flo'. What have you found?"

"What do you make of this? Although this departs a little from my firm's research topic, doesn't it seem worth investigating further, separately? Do you have any idea what the purpose might be, of altering the pages?"

"Hmm. Maybe wanting to give an impression that the state is much more right-leaning than it is? But that doesn't make sense. Everyone knew that these were just tests. No one on the outside would care about anything except whether they matched up or not."

"And you said they did."

"Yes, well, except for the station with the two retired schoolteachers."

"How much off there?"

Perrino breathed in deeply. "Seven votes. Within the margin of human error."

Latanya nodded. "I understand you didn't compare to the video."

Perrino shook her head. "It was late, my people had other commitments to attend to. Some had to vote themselves. As you can see, checking the video would have been taken another ten hours," she continued, with what Latanya perceived to be a defensive edge to her voice.

Latanya nodded. A thought so frightening began to take form that she preferred to suppress it right then and there. She suddenly felt she couldn't trust anyone. She saw not a youngish parallel test manager standing in front of her, but a Nikita-like double agent of some shapeless, all-encompassing and ruthless organization. To prevent giving Perrino the impression she was suspicious, Latanya smiled. "Okay. Maybe just some right-wing wacko, who didn't get the message. That certainly wouldn't be the first time, right?"

Perrino grinned slightly, nodding.

Latanya left, agreeing that 'Florence' would be calling if she had any more questions.

In the car on the way back, Latanya turned on Alicia Keys' "New York," as her mind wandered. She didn't want to think about what she had just seen. Instead, she remembered her trip to that city when she was eight. How her dad had bought her cotton candy after they had gone to the petting zoo. It tasted so good, and she felt so special, being surrounded by tall trees and big buildings. "And the people livin' here—" he said, pointing to the tall, tall buildings on Madison Avenue, "'dey run the country. So ya gotta be nice to 'em."

But against this backdrop a thought continued to take shape, like the realization that someone close to you has done something terribly evil, when you hadn't believed the initial accusation. If the purpose of the whiting-out wasn't to prove, to themselves, that the machine was or was not pre-programmed—a cynical, but fair, test, then those who manipulated the sheets had a far more sinister reason. A cold chill descended upon Latanya's spine.

To evade the thought, she took out her phone and called Dr. Artois. Artois had to be on set in half-an-hour, so she couldn't meet.

But the thought returned, like an oppressive thundercloud. *Could it be that they had to white-out and manipulate the voting ballots,*

266

because the machine had been pre-programmed to deliver a certain percentage of votes, Democrat or Republican? "Calm down, girl, calm down. Don't rush to conclusions," she told herself.

***

Olson was watching the Steelers game with Finnegan, when he got the call. Latanya had been seen on 6th Street. "Odd part of town for her to frequent, isn't it?"

"Not my choice of neighborhood, that's for sure."

"Good find, Agnelli. Have that corner staked out for a couple of days."

"Tonight?"

The chief had been railing about costs. Olson pondered the cost of placing someone there at the moment and weighed that against the possibility that tonight was the night a bed-ridden suspect would be crated somewhere else, six weeks after being moved there. "You just have the neighborhood, not the building, right?"

"Yeah, pretty much."

"So, she could've been visiting a relative, a student or a hairdresser for all we know."

After a silence, Agnelli responded: "Yeah."

"Fine, put Sweeney on the place starting Wednesday, when the Heinzmann job winds down. No sense in clocking up overtime."

***

Jenkins looked on as Bueller connected the wires to the metal device he had placed in the wall cavity next to the side of the door. Bueller closed the wall back up with the square piece of gypsum board.

Jenkins held the board, while Bueller began applying the instant putty to the cracks, starting with the bottom. He finished puttying the bottom crack, then started the left side.

Jenkins lifted his hand and the board stayed in place. "I'm off. Be waitin' for you in the car," Jenkins murmured, throwing the hood over his head. He hobbled out.

267

Twenty seconds later, the the middle apartment rear door swung open. A young black man six-and-a-half-feet tall held a handgun two feet from Bueller's right cheek.

Bueller opened his eyes wide. He dropped the pliers, which hit the floor with a clatter. He wasn't used to people larger than himself holding guns to his head. Slowly, he raised his hands.

"What the fuck you doin' here?!" the black giant asked.

"I'm-- Mr. Seidmiller asked me to fix an electrical spark problem."

"What problem? No one told us 'bout no problem. And what fuckin' electrician works on a Sunday. What, you ain't got no unions?!" the man barked, jamming the gun even closer to Bueller's cheek.

"It's a recaminant static spark," he said, putting together words that came to mind, hoping the crazed man wouldn't know these terms any more than he did. "Wires rubbing together. Could start a fire at any moment—"

"Shuddup," the giant said, shaking the pistol. "Keep workin'. Tell anyone ah pulled this thing on you, gonna blow yo' brains out wit' it."

Bueller nodded slowly, glancing behind the man.

Suddenly the black giant smiled. "Aw dude, sorry. I got a baby asleep inside. Just got nervous. This 'hood ain't that safe . . . We cool?"

Bueller nodded. "Yeah, we're cool."

The man disappeared inside the apartment.

After he finished, Bueller packed everything quietly and quickly into the toolbox.

He got to the street, threw the toolbox in the trunk and sat in driver's seat of the beat-up 1994 Silverado. "A coon came out from the flat. Held a gun to my head."

Jenkins shook his head. "Friendly out here. Well, with the package you just installed, that uppity nigger won't be uppity much longer, will he?"

Bueller shook his head. "I also caught a glimpse of the flat but I didn't see a nurse . . . or the reporter."

Jenkins shrugged his shoulders. "We know they're here."

"Yeah."

"Then no worries."

Bueller shook his head. "Nope."

"Good . . . you know, I was just thinkin' about that girlfriend that dumped him."

"Yeah, I remember something about that."

"What if she didn't break up with him after all?"

"Hunh?"

"Cuz' I just can't figure why a drug dealer'd be innerested in harborin' a fugitive. They don't do that shit for their best clients."

"Yeah, I guess you're right."

Jenkins scratched his head. "Naw." He shook his head. "I heard reporters can get any pussy they want. She was probably a blonde gymnast with a scrumptious set a' knockers." He gestured with his hands how 'scrumptious' he meant.

Bueller grinned. "Oh baby, oh baby."

"She wouldn't know any drug dealers or anyone from this 'hood." Jenkins waved his hand in dismissal. "Let's go, 'fore someone thinks we're here for the meth."

41

On Monday morning at 9:30, Latanya felt a sense of relief that she'd decided to render Paul back to the police. Abducting and harboring him was just wrong on so many levels.

She walked into the busy Seven Eleven and headed to the nuts aisle. She felt in the usual spot at the back of the shelf. No paper. *Strange.* She didn't know whether to go or not. As per her understanding with Michael, if there was no paper, that meant trouble and she shouldn't appear. And he always left notes early on Mondays. After buying a ginger ale she walked out.

Latanya decided that no news was good news. Perhaps Michael had just forgotten; he was never a stickler for precision when a sweet young thing got in the away.

She'd just drop in on Paul impromptu today. Who knew what tidbit of that November day's events his memory, like a rising tide, would deposit on the shores of common knowledge. She needed just two or three more pieces of information and she'd also have a report for Fauntleroy.

Latanya was going to be up at Penn for a few days, so she decided she would surprise Paul with a Breakfast Mac and an Oreo McFlurry before she left Harrisburg, despite the agreement she had made with Michael about no impromptu visits. The sun, shining between the high white clouds, made her even cheerier.

Walking back to the car from the McDonald's, she heard an ambulance siren from a little distance. It grew louder. Another siren joined in, this one a much lower tone. *A fire truck.* A second lower-toned siren joined the medley.

As she drove down the cross street to the hideout, two more police cars raced by, sirens flashing. They turned right at the corner.

Suddenly, Latanya felt goose bumps all over. As she turned onto the street, she saw two fire trucks, three ambulances and several police cars blocking that forty-yard stretch of the street from both sides. A huge cloud of smoke billowed from the apartment building windows. The police had placed barriers twenty feet from the fire trucks, on both sides.

She stopped the car in back of four others in the eastbound lane, forty feet from the police barricade. She held her hand over her mouth as they carried a body on top of a cart.

The body was covered, but still smoking. An arm, charred black, dangled to the side. Wisps of steam rose from the limb towards the fire ladder, now extended to the third floor.

She couldn't suppress a scream, emanating from the depths of her soul. Tears fell on the steering wheel and the dashboard.

Finally, after some time, she looked up through her windshield. Directly in front of her, an elderly woman kneeled down on the sidewalk, her face buried in her hands. Her body convulsed violently with each sob.

Latanya hit the steering wheel with both hands, then grasped blindly for the door handle, as if trapped in a car that just fallen off a bridge into a deep cold river. "Oh my god, Oh my *GOD!*"

The door finally opened and she stumbled out. The air, thick with smoke, smelled of burnt plastic and something sour she couldn't define. Temporarily deafened by screams, mingled with a honking horn and the roar of floorboards and walls being consumed by flames, Latanya now perceived only a haze. The heat singed her forearm hairs, but she barely noticed it.

She plodded to the nearest policeman, who was guarding the barrier. "*Is anyone alive?!*"

The police officer held his arm out, keeping her from proceeding further. "Stand back! Stand back!"

Latanya came to and stood her ground. The officer looked at her for a second. "Not sure, can't comment. These lab explosions are," he said, shaking his head. "Bad."

Once again, she held her hand to her mouth, as she watched a fireman douse the second-floor apartments of the three-story building. Smoke billowed out, like a miniature grey twister, from one of the second floor apartment windows.

One of the third floor windows exploded with a 'whoom' and dark smoke poured out of that space as well. A fireman ran out of the house, carrying another charred, steaming carcass. Latanya, now surrounded by ten other onlookers, couldn't help herself any longer. She crouched down. The tears gushed out, she couldn't stem them. The odd sour odor wafting in between the burnt wood and plastic, she realized, was that of burning flesh.

She felt someone bending down towards her. "Now, now, honey, ya can't be sure it was one of yours," an older woman's voice told her.

She kept crying for another minute, before looking up at the kindly lady and nodding. Slowly she raised herself up. Every muscle in her leg hurt at once. She wiped her eyes, and stood, frozen in place, watching the flames engulf the roof. She slowly scanned across the barriers and the ambulances.

Suddenly, she noticed something that prompted her to look again, on the far side of the other barricade. About forty yards away, a black car door was ajar. The car was large, polished, menacing. Both a foot and a crutch extended past the door, midway to the ground.

Latanya curled her hands into fists. The nails dug painfully into her palms as she quietly walked thirty feet towards the sidewalk on the same side of the street as the burning building.

She melted into the crowd of twenty people watching the spectacle from that vantage point. Between the torsos in the crowd, she focused laser-like on the black SUV. Through the tinted door window of that vehicle, in two-thirds profile she saw the blonde guy in the picture, no longer with a bandaged nose. *Jenkins, Arnold Jenkins.*

Latanya quickly looked down on the ground, and keeping her head down, she walked back to her car, her hands still clasped in tight fists. She got in and pressed herself against the back of the seat, looking on as the medics lowered the black zippered bag from the gurney onto the ground.

Unable to feel her arms or legs, she drove one block in reverse gear and pulled over.

What to do next? She let out a sob, but controlled herself enough to pull out her cellphone. Her fingers shaking, she speed-dialed a number and waited.

"Hi Mom, has Mike called you . . . lately?"

"No honey, I haven't heard from him in weeks. What happened?"

She bit her lip.

"Child, what's wrong?"

Latanya breathed in. She couldn't stop the sob that came out. She wanted so badly to tell all, that it hurt. But she just couldn't.

"A really tough day today, Mom. Had a fight with the director. I thought I might have to, have to start the dissertation over."

"Well, well sweetie," her Mom said in that comforting tone only a mother could perfect, "who hasn't had a bad moment with the boss. As long as you come back the next day with an apology and a reason, you'll be okay . . . "

"Yeah, I guess you're right," she answered, her breath breaking in two. She wiped the tears away with her palm. "Thanks, Mom."

"But honey, what's your brother got to do with it?"

"N-nothing." Latanya shut the phone. Once again she let go a sob from the very depth of her soul.

273

42

Olson tapped his fingers on the desk as if they were hammering penitentiary stone. "Well, if they were there, we'll identify them through their teeth, bone marrow. Thanks, Ted. Good you called the firemen right away. Saved the building next door, apparently," Olson finished.

"Anything else?" Officer Smith asked.

"No, not for now."

The officer left. Olson looked at Finnegan.

"Any ideas?"

"DNA'll be in by tomorrow, day after at the latest."

"Yeah, and we have Forrester's on record?"

"Michael Forrester's, sure."

"Well, if he *is* involved and he *was* in there, *and* we can save his teeth from the fire, it'll be a familial match . . .."

"Speaking of which, Phil reported that Mike hasn't been in the usual spot lately. But, then, he'd been there less and less over the past year, too."

Olson shrugged his shoulders. As unfortunate as it was for the inhabitants of the building, the fire had positive consequences. "If they weren't in that building, then one less for us to investigate. If they had been in the building, then…case closed. Mike is just a person of interest at this point."

***

The only thing she could do was wait. She felt paralyzed. She paced her apartment nervously. She sat down, taking her head in her hands. *But who could survive that?*

An hour and a box of tissues later, she decided to hope that maybe, just maybe, they had stepped out. Michael had said Paul was walking a little.

274

In any case, she had to gather all the information she had to-date, for Uncle Charlie to take to the police. It was clear who had started the fire.

She took out three pieces of blank paper, sat down and started drawing out a scheme, based on what Paul had told her:
-'Programmer's apt. --------   12:30
        -Some 'counselor' dead. Fight with bad guy
and break his arm
       -Paul and programmer leave------ 1:00

Latanya continued writing for another two minutes before she put the pen down. Out of nowhere, she remembered their summer weekend trips to Ligonier. Leisurely strolls in the fort, reflections on history in the gelato shop, runs through the tall grass, California wines and wild lovemaking in the afternoons, his musky smell . . . *Why me? Why me? Why . . . my poor Paul?*

When her tear ducts were dry and she could cry no more, she fell back on the couch, staring at the ceiling. *Could Elaine know something.* Latanya slowly lifted her arm and picked up the phone. They agreed to meet in two hours at the Bookstore Cafe.

They ordered coffees at the front desk of the immense marble-floored space and settled into armchairs on the second floor, clear across the vast room from the two only other clients.

"Latanya, I can't begin to think how you're feeling, not knowing where he is . . . or if he's even alive," Dr. Artois began.

Latanya nodded, resignedly. "I think, I think Paul . . ." she lowered her head, wiping away a tear, "might be dead."

Elaine rose from her chair, sat on the armrest of Latanya's chair and hugged Latanya. She stroked her back.

Latanya felt better from the contact. The burden of secrecy felt lighter. After a few minutes, she felt sufficiently emboldened to resume talking.

"But he was recovering, wasn't he?" Elaine asked.

Latanya nodded. *No one can know the truth.*

Elaine sat back down in her chair. "I'm just curious, though, Latanya. Did he contact you?"

275

To the last question, Latanya popped her head up inadvertently. "N-no."

"Latanya, he's a fugitive, a suspect on the lam." Elaine jerked back in her chair, and cocked her head to the side. "Have you been hiding him?"

Latanya shook her head furiously. *How did I give it away? No, Elaine's just super smart.*

"They have proof of pedophilia—" Elaine continued.

"Stop!"

Persons at the two tables twenty feet away turned to them. She lowered her head. "Just stop . . . I'll explain everything," she quietly continued.

Latanya proceeded to divulge everything she had gleaned from Paul, saying she had received the information from him during the forty-eight hour period since he had opened his eyes at the hospital, and she'd been mulling it over since then. She also mentioned the photos.

Elaine listened with intensity; she was focused as Latanya had never seen her focused before.

When Latanya finished, Elaine turned away and stared blankly through the large plate glass window. "Wow . . . wow. I, I find it all . . . overwhelming." She shook her head. "What are you going to do next?"

"Elaine, what's a fast-acting poison, that leaves no trace?"

Elaine laughed nervously.

Latanya maintained a steely expression.

Elaine stopped laughing. She dropped her smile. "Seriously? No, Latanya, don't. You will go to jail."

"But avenged."

Elaine lifted up her hand. "Latanya, tell me instead—what can I do to help this investigation through the courts. To absolve Paul of guilt. He would have wanted that so much more instead, for his family."

Latanya nodded knowingly, then sipped her coffee slowly, silently.

"What about the machines? Did you find anything out?" Elaine asked.

"Nothing definitive so far."

276

Elaine leaned in to Latanya. "By the way, did you know about the journalist and the farmer?"

"Know what?"

"Well, Dave was taken by that conversation at the shooting range, you know that Sunday, and the next day came across an article in the Altoona newspaper about someone who thought the machine had changed his vote. Dave phoned Paul late in the morning about it."

Latanya dropped her cup, clanking the saucer. "You're kidding."

Elaine shook her head. "You know how men communicate. Dave told me about this just two weeks back. I thought you'd know."

"That . . . that may be where he went between four and six," Latanya concluded, excitedly. "Do you have anything else, Elaine? Any messages he may have sent anonymously, anything?"

"Yes. He needed a psychiatric counselor to help him communicate with the programmer guy, so I found him one." That's all I remember that day.

"That's the dead guy they found in the programmer's apartment?"

"Yes."

Latanya looked up, dazed.

"Wait, there was one more thing. I got two messages from some lawyer. But I'm sure it was a wrong number."

"Wait, Elaine, didn't they find a lawyer shot to death, whose car Paul had taken, not far from the Sideling Hill tunnel?"

Elaine nodded. "I deleted them—I think one had an attachment, but it said, it said something about a murder."

"Deleted them?!"

Elaine waved her hand. "Listen, don't worry. I'll go ask for files of that number."

A man settled into an armchair not far away and returned their gazes when they looked.

"Thanks. That would be great, Elaine . . . along with the info about the poison," Latanya murmured.

"No way, Latanya."

277

"You know, Elaine, it's at times like this, I think we're all ants. And a few men in society thinks they're the collective 'man', and so they move the ants around, maybe step on a couple thousand of 'em. And, even though, all together, we far outnumber those men, we keep living in holes, oblivious to the men's power and insidiously evil intentions."

"I'm sorry, Latanya, please understand, I still can't give you that information. And . . . and if Paul's dead, would it really help anything?"

"But what if . . . what if he's alive? Maybe this would save his life." Latanya wanted so much to believe that some miracle had saved him and that the charred, distended bodies they were rolling out of the tenement building, were only the drug dealers and their forsaken clients. But she bit her tongue.

Elaine shook her head slightly. "Hippocratic oath."

Latanya looked down and smiled. "Then goodbye, Elaine. Thanks for the info." She got up and calmly walked down the stairs.

Even before the door of the retro bookstore shut behind her, she knew that Doctor Artois was right. There was no sense in the poison. She'd better just focus on starting to clear Paul's name instead. Not ten steps later, her hands and legs felt entirely numb, as if they belonged to some stranger walking across the street; she had just lost the most important man in her entire adult life.

***

By seven o'clock, the Greenburg forensics lab had found just one perp's profile in the NDIS database. Not a nice guy, either. It was an armed robber, convicted of rape twice, heroin peddling. A second had a similar profile, but had started even earlier, and participant in two deadly shootouts. For the third, nothing.

Olson turned to Finnegan. "How many more?"

"Three."

"Hmm. It would be Robinson and a nurse as minimum, 'cause looking at these guys, only nursing they'd do'd be for their pit bulls."

278

"Dr. Grady had an appointment tonight. He said it would take about five days more to get to the tests and twenty-four hours for the P-C-R to replicate enough DNA for us to gather from bodies four and five. Then two weeks to get the data back. They've got a big backlog, you know."

"Yeah, shit. Did people witness anyone escaping?"

"No. But you'll never guess what the crew found."

"What, Jimmy Hoffa? Don't keep me in goddamn suspense!" He slammed his palm on the solid wood desk.

"Glass bottles and an IV stand for suspending fluids."

"Did they analyze for melted plastics or latex?" Olson shot back.

"Yes. It appears that John Doe number five had some on the arm."

43

Latanya felt like someone had been pummeling her chest and shoulders non-stop for a month. It was only nine p.m., but she wanted to crawl into a corner and expire, a mortally-wounded bird. Her parallel thoughts, one with Paul alive and the other with him dead were still wrestling, but the thought of Paul being dead was now taking over. She finally fell asleep some hours later.

At four in the morning, Latanya awoke, her throat like sandpaper in a desert storm. She got up and put on her clothes. She had to keep moving, searching.

As she approached the Seven-Eleven, she decided to get something to quench her thirst.. Her feet felt leaden, the fluorescent lights above, blinding. She picked up a couple of colas and slowly made her way back to the front. But before reaching the register, not knowing why, she turned on her heels and proceeded to the nut section.

She stopped in front of the section and stared at the sunflower seeds. She swallowed painfully and thrust her hand toward the back of the usual shelf. Between the corn nuts and peanuts, she felt something different. Not a fallen bag of nuts or seeds, nor a smoothly folded piece of paper but a scrunched up piece of tissue.

Did I miss it this morning . . . this afternoon? No, she was sure she hadn't. She grasped the paper and slowly pulled it out. The tissue was smudged black.

She opened it.

The lettering was wavy, but legible. 'Shit go down. Boom,' the note read. *Michael's alive.* 'BF safe. B leavn in 2 dze. Ex-HRBG. 4 wk. NALN 10.5' *Leaving in two days?* Then the standard end – 'Need anything, leave a note by 10:30.'

Latanya burst into tears. Just as suddenly, she threw back her head and started laughing. She turned around and saw the clerk looking at her, his head bent sideways, as if at a person on the very edge of sanity. It didn't matter. She laughed again, giddy with joy.

After another half minute, she regained her composure. *Damn right I need something.* But rather than write something right away, she knew she would have to weigh everything first. She'd drop it off by mid-morning.

An hour later, she hit on a plan. She started by sending an SMS to Elaine. 'I have things that I know Paul would have wanted you to have. Could we meet today at 7:30 a.m. at the Espresso place across from Capitol'.

At 7 in the morning, Latanya's phone rang.

"Oh, Latanya," Elaine started with an empathetic tone.

"Let's not drag this out, Elaine . . . I felt something, something go off. It told me—don't wait any more for him. He's, he's gone." Latanya forced herself to sob. "The sooner I get this over with, the better . . . and he admired you so much."

"Sure Latanya, sure. At seven-thirty, across from the Capitol."

Latanya hung up. She had one more thing to do.

At precisely 7:28, Latanya pulled up on Third Street. Straight ahead, the sun rays had just alighted on the gleaming greenish-yellow Capitol dome. She walked up to the counter of Into Coffee and ordered a cappuccino. As she waited for Clayton, the ear-ringed barista to serve it up, she turned around and saw two men walk through the door. One of them, the taller of the two, had a long scar on his left cheek. He didn't greet Clayton. *He's not a regular . . . but he looks familiar.*

She took her cappuccino and made sure to sit at the high table in a corner closest to the window, looking out the window again. There were few things as lovely in Harrisburg, she thought, as the Capitol building, crowned by the gold-leafed 'Commonwealth.'

The two men sat at the adjacent table. Latanya always thought social norms, perhaps unspoken but universally accepted nonetheless, dictated that in the absence of other customers, one didn't sit next to the occupied table if a more distant one was available. *Strange.*

Elaine walked through the door. She glimpsed Latanya in the corner and waved.

281

Latanya nodded back, as she pondered what she could divulge here.

Another strange thing. The two men weren't talking to each other. Why else goes to a coffee shop two-at-a-time?

As Elaine joined her at the table, Latanya decided to take a conversational detour. She reflected on the dome. They talked about old times. She burst into tears several times, and each time Elaine comforted her with a warm, sincere hug.

Latanya handed a large shoebox to Elaine. "Everything he would've wanted you to have is here."

"Elaine, I need some fresh air. Can we go outside for a short walk?"

"Sure."

Latanya glanced past Elaine at the two men. The taller of the two started folding up the paper and drank the last of his coffee.

They exited and walked down the street. Latanya felt the tall man was following, some twenty steps behind them.

As they passed the gray-stoned Presbyterian Church, Latanya stopped and turned to the doctor. "Elaine, I'm not extremely religious, but I feel I, I *must* . . ." a tear fell down her cheek. She wiped it away with the back of her hand. She looked out the corner of her eye and saw the man stoop down to grab a real estate-listing throw-away from the scuffed-up plastic dispenser. "Will you come in with me?"

Elaine shrugged, then nodded.

They opened the massive doors and stepped in. Just before the door closed completely behind her, Latanya glanced back and saw the man walking past the church, eyeing her feet.

The two women took their places in the next-to-last pew. The church was empty. Latanya put her hands together in prayer and lowered her head.

"Elaine," she whispered, "Paul's alive."

"Now, now Latanya. It must be very, very hard for—"

"No, really. I got word from his guardian last night."

Elaine inhaled deeply. "Latanya, his spirit—"

"Elaine!" she uttered, loudly enough to hear echoes off the stone walls. "You were right. I've been involved . . . I know who's been harboring him. He reported to me last night."

"Oh my goodness."

"Yes, and now I ask forgiveness, Oh Lord," Latanya responded loudly. "It'll hit the papers today, in two days at the latest," she whispered. "Not about Paul, but about the identities of the others. As soon as they know, they'll be after him again. Help me . . . please."

Elaine opened her eyes wide. "Latanya . . . but how?"

"What can I use? Will arsenic work?"

"No. You just . . . can't."

"Listen Elaine. Yesterday you were right. If it were just revenge, it would be stupid." Latanya dropped her head as if to emulate even deeper prayer. "But it isn't revenge now, it's protection. I saw who organized the bombing."

"What?"

"This blonde guy. And that guy in the coffee shop who was listening to us I've seen somewhere before, too—" A chaplain emerged from the rear of the church. He walked briskly towards them down the aisle.

"Oh Lord forgive us our trespasses as we forgive those . . ." Latanya loudly recited as the priest approached to within twenty feet. She continued, head bowed, with her prayer.

As he passed and walked towards the entrance, she leaned into Elaine. "That guy in the coffee shop. I think he's part of their group. That building explosion—that was no accident. They were after Paul. Please Elaine, I need your help." Latanya explained in detail what she had learned about Jenkins and about the montaged photos.

Elaine shook her head. She had a dazed look in her eyes. "The police should—"

"Pull out all stops to protect a pedophile fugitive?" Latanya shook her head. "Paul needs your help . . . Listen Elaine, after our talk yesterday, I looked some things up. The guy is a player."

"Okay, okay. What you want is . . . but I can't get it for you."

"I'll get it, if it's a drug, a syringe . . . anything. I'll get it."

283

Elaine bit her tongue. She turned straight ahead and bent her head down. "Lord, forgive me for what I'm about to do." She turned back to Latanya. "You didn't hear it from me, but what you want is strychnine. It can be ingested accidentally. If the guy eats mushrooms, that'll work. It'll take a day, maybe two."

"I'm leaning towards superfast."

The chaplain walked past again.

Elaine bowed her head down. She lifted it up as he walked in back and towards the southeast corner of the church. "You say he's a player. Any chance you could get intimate with him?"

"Hmm. Hadn't thought of it, but if I have to . . ."

"Listen, if you can get him to take Viliagra and another pill, then the possibilities open up."

Latanya nodded.

"As a back-up, there's potassium cyanide. But you'll need all kinds of protective measures . . ."

"Like what?"

"Rubber gloves, not latex ones."

"Where do I apply it?"

"Anywhere on the skin. The more you apply, the faster it works."

"How fast?"

"Anywhere from three to fifteen minutes. Three if you get it in his mouth . . . But whatever you do, you can't even inhale it, let alone touch it. Plus it can be traced, so not the first choice."

"Okay. What about date rape drugs?"

"Ketamine, G-H-B, Ritalin."

"I need one that totally knocks out fast. Just thinking off the top of my head, but would there be one that starts working in fifteen minutes and for two hours delivers no memory recall?"

"Hmm. That would be 'G-H-B.' On the market, you'd call it Georgia Home Boy. Like being knocked out, but still functioning. You create a zombie that listens to you. But for fast and fatal?" Elaine cradled her chin in her right hand, and reflected. "One could go with Viliagra and a recreational drug, an amyl nitrate."

"What the heck is that?" Latanya asked.

The door opened, and she heard someone stepping inside. As he approached, she lowered her head. "Dear Lord, please grant

284

me the strength to overcome," she said, at normal volume. As he passed them and sat three benches in front of them, she once again turned to Elaine. "What else about them?" she whispered.

As she waited for Elaine's response, she recognized the clothes of the man sitting down – it was other of the two men at the coffee shop, the shorter one.

Latanya leaned in to the doctor. "Wait, Elaine. Write it down." She raised her head back up and faced the front, clasping her two hands in prayer, as the man turned around and glanced at them.

Elaine jotted some things down.

"Thank you, Lord," Latanya loudly recited. "Shall we go?" she asked, loudly, wiping tears from her eyes.

"Yes. I'm so sorry for your loss," Elaine responded, loudly enough for the man to hear.

They got up, turned around and left.

On the way back to her apartment, Latanya stopped in a diner, strolled to the back and sat in a corner table facing the door. No one had followed her.

She took out a small piece of paper and started composing her request. She finished with the phrase: 'Ready 2 mt any tm.' On her way home, she went to the Seven-Eleven and placed the note underneath the top shelf in the potato-chip section, as per agreed-upon procedure.

Two hours later, she returned. The note read, 'Mt me back-up p. dozen.'

At twelve o'clock, she showed up at Happy Dee's restaurant. It was a respectable sandwich joint and full at this time of day. She scanned the tables, but didn't see her brother. She turned around slowly, preparing to leave, when a gray-haired, gray bearded man, sitting just to her left, rose halfway and said: "Miss . . . hmm, miss?"

The scraggy voice didn't fool Latanya. But she barely recognized Michael beneath the make-up and gray hairs. She walked up to him.

"Miss, could you help me read what's in the menu? Ma eyes ain't what they used to be."

"Sure." She was overwhelmed by the smell of smoke, of burned wood and plastic, though he had, no doubt, cleaned himself and bought new clothes.

"Mighty nice o' you helpin' a' old man like me."

Latanya bit her tongue. She wanted to laugh. And not just from the comic aura her brother was now exuding. She wanted to laugh out of joy, to hug her brother tightly, like she used to do when he was five and she was seven. Instead, she clamped down inside and assumed her best poker face.

"Would you mind takin' a seat and readin' the whole menu, miss?'"

She looked around, as if embarrassed.

"Oh, all right." She sat down and picked up the menu. "Appetizers. Got salad . . ." she read off at normal volume. "So what happened?" she whispered. "Chicken fingers . . ." she resumed reading.

"They blew up the dudes next door," he replied in a mumble, before lifting his head up. "Just da sandwiches, please, dis old man ain't goin' for no appetizers."

"You smell like a barbecue gone funky. How'd you get out?" Latanya whispered. "Pastrami on rye?" she asked loudly.

"We ran out the back hallway 'bout a minute after the explosion. The wall saved us," he whispered back. "Stop right there. Pastrami on rye sounds fine," he answered back normally.

She nodded. "Those guys were what, makin' meth?"

"Meth, Crack. They were cookers. I waited every morning until things quieted down before stepping out. So anyway, whatchu need?" Michael murmured.

Latanya took out her note and pressed it into his left hand, under the table.

He looked down at the list he held below the table. "Sis, what you be askin' fo', a sex-death cocktail? Forget-about-it."

"Mike, listen . . ." she whispered back.

"Yeah Miss, side a' slaw would be nice," he answered, loudly.

"I saw the blonde guy with the crutches."

Michael's expression turned to pure cynicism as he bobbed his head, hindi-style. "Again? You're obsessed."

"Michael, I saw him sitting in a car, watching the ambulance crew carry out the burned bodies."

Michael was silent. "You one-hundred percent sure it was him?"

She nodded. "Crutches and all," she said quietly.

Michael rocked back into his seat. He rubbed his jaw hard with his thumb and forefinger. "Shit, sis, that changes everythin'. . ." He shook his head. "We've gotta get outta here. Got to get out far from here. before they find out we're not dead."

The waitress walked up. Latanya communicated their order and the waitress left.

"This morning, I tried to throw them off course. One was following me and I pretended I thought Paul was dead. Big act. Went into a church," she continued.

"Church? Man, I gotta tell Momma about that."

"You best not!" Latanya retorted. She felt giddy with relief, she so relished being able to jest with her little bro', very much alive.

"So the poppers, Viliagra, G-B-H, that's no problem. But where do you suggest I get this cy-o-nide?" he asked.

"Paul's doctor friend said you can get the cyanide in a jewelry store."

"Like that's where I hang."

Latanya looked at him seriously.

"Okay, I'll figure it out—" Michael said, scratching his gray beard. "Man, this itches."

"And it don't smell too good, neither, bro'."

"Ha-ha. Funny. Listen, Sis, this stuff about this dude and these guys following you . . . that's ba-a-ad," Mike said, as he fingered a toothpick in a plastic wrapper. "I have to take your Paul away. If we got the Bureau on our tails, or some execution squad, we've gotta clear out of Harrisburg by tomorrow morning."

Latanya wanted to burst out in tears. The constant running, the uncertainty. With a feeling of resignation, she nodded. "But I need these things, Michael. And the name on the driver's license, make it Elizabeth Matthews or something, you know, assimilated."

He nodded. "Are you sure—"

287

"If I don't stop him, he'll find you again," she interrupted. "The police didn't know where you guys were. That asshole did."

Michael looked blankly ahead. "Sis, it's a human life."

"Did you never—" she lay her hand on his forearm ". . . in your line of work?"

Her brother looked down at the table, resigned. "Just don't you get in any trouble. In fact, I can't let you see your man before we go."

Latanya pulled back her head. "I have to."

Michael looked her firmly in the eye. "If you do, it'll be the last time ever. Come on — they were tailing you."

Latanya gritted her teeth. She hated to admit it, but he was right. "Okay. Just get the stuff to me. I'll take it from there. Oh, and we don't have time for notes now, so just call me on this new phone number, from a payphone."

"Yeah, like you can still find 'em around."

The waitress approached. He cleared his throat loudly and fixed his gaze on Latanya. "Whatchu thinkin' young lady. I gonna be yo' sugar daddy?"

44

Latanya half-listened to Jennifer, her most dedicated student, as she explained her progress on her mid-term project.

"So I have information on the socio-economic backgrounds of the Democratic students, but not on the Republican ones."

"When will you have those?" Latanya probed.

"It'll take two weeks."

"Two weeks?"

"They are all preparing to attend G-PAC next week in Washington."

"And would their officers also be going?" Latanya asked calmly, hoping that Jennifer hadn't detected anything in her previous overreaction. Delays were a constant in thesis work.

"Everyone. Including that Corn Cob Caucus representative. They said he's going to be hosting an information session there."

"Well, I guess you're right, you'll just have to wait until after that. I'll extend your deadline by twenty days, but not more, okay?"

Jennifer nodded, smiling.

Right after the meeting, Latanya stepped into the bracing cold. Her plan was coming to fruition.

The Forensics faculty was a hundred yards away from her department, conveniently close. Dr. McKinley was sitting in his office, glasses perched on his nose. She knocked on the door.

"Oh, Latanya, hello!" he belted out, putting down the paper he was reading.

"Hi, Professor McKinley."

"Thought you'd forgotten about me. I heard about your boyfriend's abduction from the hospital. I'm sorry."

Latanya nodded glumly.

"Any word yet on his whereabouts?"

"None," she said, pleased that she didn't have to lie. Besides—either in Philly or in one-of-five-states bordering Pennsylvania—she could not have told him where Paul was.

Mckinley nodded.

"But you're still building his case?"

"Yeah, because I want to clear him . . . and I still want to believe he's alive."

McKinley shrugged his shoulders. "So, what can I help you with this time?"

"Just a procedural issue."

"Go ahead."

"I'm just wondering. The Pennsylvania police and State Troopers haven't found him in over two months since his abduction."

"Yes?"

"And assuming he's alive . . . I know this may be wild optimism on my part, but assuming he might be alive, Professor Mckinley, what happens if he's been taken to another state?"

"Well, in that case, and I'm sure Pennsylvania's done this, they'll send out alerts to all police departments in the country. There's a law of rendition, or extradition."

"And how does that work?"

"Well, they'll ask the police departments to look for the suspects. Then, if they find them, they'll send them to the state asking to extradite them."

"And how fast will the other police departments react?"

"Well, the truth is, not very. I mean, sure, if the suspects look and act like the members of KISS during a performance, then they'll be apprehended swiftly. If not, a suspect could escape capture for months, years even. Obviously, no other state's law enforcers have the same incentive to catch someone," McKinley picked up his pipe. "Serial killers, perhaps, excepted. But that's assuming they kill in the state where they're hiding. Then the FBI becomes involved. But if they're just fugitives from one state. . ."

Latanya nodded. "You and I know he's not, but what about pedophiles? Does that change anything?"

McKinley twirled his pipe. "Naw, not really. Unless you were to add 'serial killers' or 'ring' to the end of the word, or if they've

290

just abducted a child. But for the remaining one of a pair, who is lying low . . . I wouldn't bet large that they'd start a dragnet for him."

***

Chief-of-Police Stuart knocked on Lieutenant Olson's door. Olson turned his chair around. "Well hello, Captain Stuart, what brings you here?"

"The usual monthly Kumbaya. Long-live interforce cooperation."

Olson guffawed. "Oh, yes, I believe that, coming from a cynic like you."

"In action, as we speak," he said, throwing a folder on the desk, "because you've been good at keeping your guys busy. Don't you have any real cases to pursue?"

"All the time. We're on this one because we just don't want any of 'em to get away."

"Anyway, a report on the fire. An eyewitness said something about an old black lady, a home boy and a white guy walking from that building right after it blew up."

"Hmm. Not your usual suspects."

"No, especially since the white guy had I.V. tubes hanging out of his clothes."

Olson rose and opened the folder. He glanced down the page, his eyes focusing on the key words – 'I.V. tubes'.

"Well, thanks Tom," he looked up. "I owe you one. A big one."

After Stuart left, Olson picked up the phone and called Finnegan.

"Finnegan, Robinson's alive and survived the bomb blast fire. It's time for a Philly stakeout sandwich. Have Michael Forrester's corner watched. And bring in that lying girlfriend straight away."

***

She had delayed the meeting until the afternoon of the following day – the longest time possible. But the delay had not brought her any real comfort. Latanya felt her skin crawling in the

291

gray police station room. The table was all scratched up, she surmised, by guilty suspects who had expressed their nervousness—or desire to be somewhere else—by doodling on it with paperclips. After two minutes, Olson walked in.

"Miss Forrester, you've been hiding something from us."

"What?"

"Don't play dumb. You've been harboring a criminal."

"Officer Olson, I have not. What makes you say so?"

"This picture of you on the corner of Radnor and 6th streets."

Latanya kept mum. Her heart pace increased and she lifted her right hand fingers to her mouth. *Don't chew now.*

"You, of course, know about the fire that took place here two days ago."

"Front-page in the papers."

"Right. So, you obviously knew where he was."

She leaned forward and lay both her hands, palms up, on the table. "What are you saying? That I know something about this house, just because I love the Jamaican jerk chicken that I buy from the darling old lady on one of the cross streets?"

Olson shook his head. "Okay. You want to play it that way? It'll come back to haunt you," Olson countered. "But I'll let that go for the next few minutes," he said, brushing something off his jacket. "Robinson escaped on his own two feet. Which means he gave you information. Since you claim you cannot lie, tell us what he said."

"Lieutenant Olson. . ." Latanya shook her head and sighed. "You gotta understand, I'm really tired of this whole thing. As a normal woman who wants to raise a family, I thought long and hard about it before I phoned you that last time, I cannot take a risk marrying a pedophile."

"We're not sure he is."

Latanya shrugged her shoulders. "Look, if you feel you need to follow me twenty-four-seven, what can I say, do it. But I'm going on a date this weekend, in Philly. It's someone my older brother set me up with. An upstanding lawyer from a reputable family, so when I go on my date, if you don't mind, I want it to stay a twosome," she said, matter-of-factly. "If you want to pursue this case about Robinson, please do it without me. Next time he contacts me, I promise to tell you." She managed the last phrase well; it sounded like she couldn't care less.

It didn't take long for Latanya to realize she was being tailed. She noted the two men and acted exactly as she would any other day. She had given Michael a two-week supply of fluids that Wednesday before he left for destination unknown. The nurse said that he was already ingesting normal food most of the time, so this was ample. In exchange, Michael had provided all the items she needed, so she planned to do exactly nothing out-of-the ordinary for the rest of the two weeks, aside from the blind date with the Philly lawyer.

Once she told him about the men tailing her, Michael decided against spending even the fifteen minutes hangin' on his usual street corner that Wednesday before leaving. He would just focus on his established clients, encouraging them to stock up with a few weeks' supply and hoping they could demonstrate a higher-than-usual degree of self-control.

Latanya looked at herself in the mirror. She looked strikingly elegant in the black sleeveless dress. The black Jimmy Choo high-heeled shoes with closed toes also cleared a very high style bar.

She spent a passable evening with the lawyer, and then confessed he wasn't her type. She'd known from some time back that those big city lawyers seldom were. They were usually smart, often charming, but at their core obnoxious. And all of them possessed a blinding sense of sexual entitlement. She didn't play that game, and this one turned out to be no different in his reaction.

Before the 'end' of the evening, he branded her selfish and wasteful of his time.

Too bad. Had he been nice, she would have forwarded his number, in good faith, to two or three of Harrisburg's most desirable African-American pin-ups. But she wasn't about to start pimping for a jerk. As they parted, she told him as much.

She couldn't help smiling, though, when she thought of the two poor suckers tailing her. They probably even questioned him afterwards. A fitting follow-up to the lawyer's parting statement;

293

"I'm pursued day and night by hotties, if you only knew." *Yes, you most certainly will be now as well . . . just of the same sex.*

She was called in again on Monday to the police station to repeat the information Paul had provided her, and told them he had not contacted her. Technically, she was not lying, he hadn't contacted her.

The smokescreen appeared to prove effective. That, along with providing her car for inspection, to make sure it hadn't been used to transport the three fugitives, was apparently enough to convince Olson that she was cooperating with them.

On Tuesday, she made an appointment at a hairstylist's in Baltimore that specialized in hair extensions.

45

Latanya watched from a distance of thirty yards, as Jenkins weaved through the tables, supposedly checking out a place to sit. In fact, he was scouting girls' cleavages. She had watched him doing this for two days. His eye flitted from one face to a chest, then another.

Latanya noted that he didn't fixate on any of them, which was good. But just in case, she was made up differently now – she had barely recognized herself in the mirror this morning.

He sat down, towards the far side of the open area lounge, at a small table, alone. The timing was perfect. Another attendee, a reasonably attractive brunette of average height, dressed in what appeared to Latanya to be an Omar de la Renza suit, sat alone at one of the tables at the edge of the dining area, far from the table at which he was seated. She was well into the middle of her meal and her friend had just left.

Latanya briskly walked up to the table. She smiled. "Hi, would you mind if I sit here."

"No, not at all, go ahead."

"My name is Elizabeth Mathesen, from Ohio," Latanya said, extending her hand enthusiastically, accompanying the gesture with a ten megawatt smile.

"Clarice Goodson, New Jersey," the diner responded, extending her arm as well, and smiling back, though less enthusiastically.

"It's inspiring isn't it?" Latanya said.

"Oh, sure is. Who have you heard so far today?"

"Well, there was Jersey Hannitous and Hirosaa Ayoliya,"

The last one is fine, good points on the shortfalls of Islam, but the first . . . Lord help us. The only thing I'd be teaching today, if this guy's views had been followed when I was growing up, would be which cotton was ripe for picking.

While the two were chatting, the waiter approached, and Latanya ordered an Everclair, the same brand Goodson was drinking.

The waiter returned, served the water, and Latanya politely asked if he couldn't come back in a few minutes, as she hadn't had time to look at the menu.

"And what did you think of Senator Berry's talk?" Latanya asked, as she poured the water to the same level as Clarice had in her glass.

Latanya didn't listen to the content of Clarice's response, but intoned positively when Clarice paused, as she focused on her task. While propping the bottom of the menu on her lap, the top of the menu overhanging the table, she broke off the top of the plastic vial containing the GHB and held the vial in her left hand, while looking at Goodson, she picked up her water glass, sipped a little from it, and brought it down to her lap. She nodded emphatically when Goodson mentioned the wholesome philosophy of the speakers. Latanya emptied the GHB into the glass, then put the glass back on the table and closed her menu.

The waiter came back and Latanya placed her order for Asian chicken salad.

"I'm amazed by the people that came here. In fact, I'm sure I saw Governor O'Donnelly from Kansas. He was just over there," Latanya said, pointing with her chin to Goodson's left.

Goodson looked over. Latanya deftly switched glasses. "He was over by the globe in the middle," she added.

"I don't see him," Goodson responded, still looking.

Latanya looked over. "Oh he must have gone, but anyway, a whole new post-democratic era is approaching, and don't you find it exciting?"

"I do-oo," Goodson responded.

Latanya smiled at the fact that Miss Goodson didn't catch the double-entendre. *Yes, enter the plutocracy, hurrah.*

"I dunno about you, but it feels hot in here," Latanya said, taking her glass and drinking down the contents.

Goodson lifted her glass and drank about a quarter of the remaining water.

They talked for another four minutes and the salad arrived. Latanya had a few forkfuls of food, which she could barely swallow. She also confirmed that Goodson had her own room; befitting a young lady in an Omar de la Renza suit.

Goodson signaled to the waiter and he promptly walked over.

"Check please, put it on my room tab."

The waiter brought the check, and Latanya focused on the largest number on the check. '5-2-1'.

"Oh Clarice, before you go, could you just write down your name?" Latanya asked, taking out her notebook and pen.

"Sure."

"It's just that there are so many wonderful people here, I just can't remember everyone."

"I so do understand you."

"Or else I'm getting senile. On that note—" Latanya leaned in to Goodson, "let me share this joke with you."

"All right," Goodson responded, smiling prettily.

"This couple, in their eighties, go to the doctor for a check-up. They show up fifteen minutes late and explain that they had checked the calendar at the start of the day, but managed to forget again in the meantime. So anyway, he examines them thoroughly, and reports they're both healthy," Latanya started.

Goodson nodded.

"But he says they should start writing things down more often, so they wouldn't forget them, even the small stuff. Well, that night they're at home, watching T.V., the husband turns to his wife, and says 'I'm going to fix myself a sandwich. You want anything?'"

Goodson nodded again, but sloppily.

"So the wife responds, 'Could you bring me a bowl of ice cream?'.

'Sure,' he replies. 'Do you think you should write that down to remember it?'"

Goodson tried to lift her bag onto her shoulder, but she couldn't seem to get the straps around her arm.

"'No, I can remember that,' he said," Latanya continued, "'So the wife says, 'I'd like some strawberries on it, too. Do you need to write that down?'"

Goodson shook her head and re-focused, as if she had nodded off but was coming back.

Latanya paused a second before continuing. "The husband replies, 'No, I can remember that, too. Ice cream with

297

strawberries,' he said, irritated by now. 'I'd like some whipped cream on it, too, the wife asked. 'Can you remember all that? The doctor said you should write things down,"

Goodson's left eyelid began drooping.

""'For goodness sakes, I can remember that. I don't need to write it down. A bowl of ice cream with strawberries and whipped cream,' he said, now really irritated," Latanya plied on, turning to her right and noting that Jenkins was still at that table and had just about finished his lunch.

Latanya continued focusing on Goodson's left eye. "So, off he went to the kitchen. About twenty minutes later he returns with a plate of bacon and eggs and puts it in front of his wife. The wife stares at it for a moment, then looks up at her husband with a puzzled expression on her face and says, 'You forgot the toast!'"

Goodson's head dropped so low she almost hit the table. If she got up now, she'd fall.

"Excuse me Clarice, but I just saw the director of the National Handgun Association, and I needed to ask him something. I'll be right back, but *wait for me*," Latanya instructed.

Clarice moved her drooping head up and down, resembling a big lazy seal.

Latanya got up and walked quickly to Jenkins table. Jenkins had just been joined by a young man, and as she approached, she heard him say, "So you agree, taxes are theft."

"Sure," responded Jenkins, "Except enough for defense."

"Excuse me, Arnold Jenkins?" Latanya asked, with a straight-laced accent. She smiled her thousand-dollar smile, hoping against hope that he wouldn't recognize her.

"Hel-lo?"

A complete blank. Perfect.

"Well I just listened to your talk this morning, Mr. Jenkins, and have to say I was just so—" she said pressing her inner left breast, pushing it out slightly. "Inspired.

"Well thank you, miss." He looked for her nametag.

The young man scratched his head. "I'll come by later, Mr. Jenkins."

"Yep you do that now, while I answer Miss—" he replied without looking at the young man, who promptly got up and left.

Latanya pulled her tag up. "Liz Mathesen, Ohio."

"Well Elizabeth, it is good to see people of your, uh . . . background, so eager to stand up for the right causes."

"Suga', after hearing you this morning," she continued, sitting down next to Jenkins, "I'd like to see something else stand up," she purred, gently brushing her hand on the inside of his thigh, "If you can be in room 'five,' 'twenty,' 'one' in exactly twenty minutes, I can... help."

Jenkins smiled back. "Babe, I think I can be a patriot."

She got up, turned around, looked over her shoulder and winked at him. She noticed a couple of younger women approaching him. She slowly walked the catwalk back to her table, keeping her back ramrod straight. *He's in the bag.*

As she approached the table Goodson's head nearly fell into the empty dessert plate. She jerked it up at the last minute.

Latanya sat down. "Clarice, how you feeling?"

"I can't, can't get up'" Goodson mumbled.

"Oh dear, must be something you ate. Let me help you." Latanya dropped twenty-five dollars for the salad and tip and grabbed her handbag. She then picked Goodson up by the shoulders.

Goodson had enough muscle control to stand up, but she was wobbly.

Latanya lifted Goodson's purse from the chair and adjusted it on her own right shoulder. She then grabbed Goodson's hip with her right hand. From the side, Latanya figured, they'd be looking like old friends just catching up with intimate details of their lives after one too many drinks. She commandeered Goodson across the vast marble floor to the bank of elevators.

During the sixty-foot walk Goodson's head rested on Latanya's shoulder.

"Where's your room key?"

"Front . . . inside . . . pocket," Goodson replied groggily.

As they arrived at the elevators, Latanya retrieved the room key card. "Now Clarice, a gentleman from the hotel will take you to your room."

Goodson nodded slowly.

Shortly, one of the hotel bellhops casually walked past them.

Latanya glanced at the tag on his shirt. "Aguilar-"

"Yes, Meess?" the stocky man responded, turning around to face her.

"Could you do me a favor? This conventioner's not feeling well. Would you please take her to her room? It's five-twenty-one."

"Five-twenty-one?" he asked, confounded.

"Yes, thank you. I think she had drunk a little too much. I have to go tell her friend, the man she was dining with there, who has her room key." she said, gesturing towards Jenkins' table, making sure the bellhop saw him.

As soon as the two entered the elevator, Latanya walked back to the bar. She saw an untouched lime on a finished drink rim, panned the bar area to be sure no one was looking, and took it. She returned to the elevator bank.

Latanya exited the elevator on the 4th floor, and walked to the stairwell. As she walked up the steps, she put her long hair up, as she had practiced a dozen times. She grasped a long hairpin and pinned the hair into a tight bun.

She stepped out and approached the hall with rooms 520-529. When someone stepped out of room 526, she took out her phone, put it to her ear and focused her eyes on the ground. She walked past the person and past room 521. Once at the end of the hallway, Latanya turned around and looked back to room 521.

Aguilar opened the door and walked out.

Latanya quickly hid around the corner.

She waited until she heard Aguilar walk to the other end of the hall, where the service elevators would be. Once he was out-of-sight, she walked briskly to 521. Using Goodson's key, she opened the door and entered. She glanced at Goodson.

Goodson was 'out'. Aguilar had taken the cover off the bed, so Goodson lay on top of the sheets, her head on the pillow. Her eyelids were lowered to reveal mere slits.

"Clarice, we're gonna have some fun now. Especially you."

"Fun now. Me," Goodson gurgled back.

Latanya hung her bag on the chair in front of the desk. After kicking off her shoes, she stripped down to her burgundy Valencia's Stash undergarments. She removed the rubber gloves

300

from her bag and put them on. Then she took the two blister packs and popped four little blue-green pills onto the table. She placed the empty blister packs back in her purse.

She then went into the bathroom and cleared the little bottles from the plastic tray. She snapped up the tray and the two glasses and returned to the room. She put the glasses on the work desk, then walked to the bed and wedged the tray between the pillow and Goodson's cheek.

As Elaine had predicted she would be, the young woman was salivating profusely.

Latanya returned to the table and using the bottom of the bathroom glass, crushed three of the blue-green tablets. She swept the powder into the glass with her left gloved hand. She stepped to the minibar and removed a cola and a rum bottle. She poured the beverages over the bluish-green powder and with her finger, stirred it. She then crushed two of the three white 'poppers' into a powder and put it into the glass. *Amyl Nitrate? I wonder what it is really used for. Doesn't matter.* She garnished the glas with the lime from downstairs.

Latanya wet a wash cloth and returned to the desk to wipe the remaining powder off. She wiped her finger as well, but left the wash cloth there. She left one 'popper' on the table. As she passed the bed, she dropped a tube of lubricating jelly on the bed.

Latanya took the tray from under Goodson's mouth and placed it on the nightstand closest to the door. She looked at Goodson's face.

The eyes had closed.

"Clarice, take off your clothes."

Goodson didn't respond. Latanya sat her up. "Get undressed to your panties and put your clothes away!"

Goodson nodded. She proceeded to take her shoes off, unbuttoned her skirt and pulled it off. She folded it neatly and put it on the shelf in the closet. The shoes she put away in the bottom shelf on the right side of the closet.

"Now Clarice, we're going to surprise someone in about five minutes. But for the surprise to work, you're going to have to hide in the closet."

"Uh-huh," Clarice said, quietly.

On the left side of the closet hung Goodson's beautiful suits, but below them was just a shoe tray. Latanya pulled it out, before lifting Goodson to a sitting position, with her head leaning against the vertical wooden plank dividing that space from the shelves. "Sorry girl, it's a little cramped, but you're gonna have to wait here a little while anyway."

Goodson mumbled unintelligibly.

Latanya shut the closet door. She glanced at her watch. Less than two minutes remained.

46

Jenkins finished his conversation with the two twenty-year-olds, who had wanted to know how to volunteer for work on a political campaign. "First thing you do, is pick the winnin' candidate," he had replied.

The coeds looked at one another, puzzled. They were cute. But these were the ones he'd coach - until he got bored, not the ones he'd hunt, which received his full attention. And one stunning African-American had just offered herself up for prey. He wondered if he should do a full background check on a certain 'Elizabeth Mathesen.' But he would eventually be asked why, and he'd have to make up another cock-and-bull story. Besides, the process would take two, maybe three hours. He had an itch now.

Instead, he had another idea. After paying his lunch bill, he got up and headed to the registration desk. He chatted up one of the receptionists and said he was looking for a friend by the name of Elizabeth Mathesen. The receptionist tapped a few keys on the computer. "Yes . . . Elizabeth Mathesen."

"From Ohio?"

"Yes, Cincinnati."

"Well that's just wonderful. What a coincidence. Whenever did she get here?"

"Late on the first day. Would you like us to leave a message?"

"No, not necessary. Ah'll find her soon enough. Thank you, Bernice." *Great, she's a legitimate attendee.*

He looked at his watch. Just enough time to go up to his room, pick up the condoms and take the stairs down one flight. *Maybe take a blue-green one and make half an afternoon of it.* After all, she looked like someone he'd seen in the movies, but he couldn't remember who.

As rode the elevator up, he let his mind wander. He was elated that someone high up in the order of things had recognized his talents, his devotion to the cause and had, subsequently, tapped him for this mission. A mission that, if executed well, would lay the foundation for a 100-year real conservative domination of the Congress and Senate, and subsequently over life in America. *The Founding Fathers never intended for the landless, women and niggers to have a vote in this Republic – so we're just rightin' a few wrongs.* The experiment up North was the first step in systematizing it, to avoid the close calls of '04.

He switched mental gears and reflected on his upcoming rendezvous. He concluded that the black race, with a few notable exceptions, was so unfit to share the reins of power, it made him giddy to know they would soon be locked out of it almost permanently. In the meantime, it was good that a few joined the party. That would keep it politically-correct for outside observers.

***

Latanya adjusted the push up bra a little, turning to the mirror. It showed two inches of cleavage, enough to communicate about 500 degrees of heat.

She returned to the desk, pulled the glove tighter on her right hand, and then picked up the second bathroom glass, along with the small vial in the self-sealing bag. She stepped to the nightstand with the tray. *God, I hope I don't have to use that.*

She removed the vial from the bag, unscrewed the enamel top and placed it on the nightstand. While holding the vial at arm's length, she carefully poured half of the clear odorless liquid, about a teaspoon's worth, onto the saliva-filled tray, quickly capping what she had just poured with an inverted bathroom glass.

Upon moving her elbow back, she knocked the enamel top of the vial off the nightstand. It rolled under the bed. *Shit.* She couldn't pour the remaining contents into the tray, or it would be too liquid. *'Keep it away from your face'* she remembered Elaine saying. She transferred the open vial to her left hand, and held it at arm's length, while she got down on her knees. With her right hand she groped under the bed.

First to the left. Nothing.

To the right. Nothing.

Then deeper, in the middle. She felt the cap.

A knock at the door.

No. Frantically, she screwed the cap onto the vial. "Be right there, honey!" she announced.

She jumped to her feet, put the vial back into the self-sealing bag, and thrust the bag back into her purse. She stripped off her left glove.

The door handle jostled up and down.

No time to put the gloves away! She bent down and threw the gloves under the bed. She stepped to the door and put her right hand on the handle. She heard a scratch coming from the closet.

The closet door swung open.

Damn. "Just a sec!"

Goodson's head and upper torso leaned out, motionless. Latanya bent down, and using all her strength, lifted Goodson and placed her further back into the closet. She shut the closet door, jammed the back of her right foot in front of it and shut off the light that beamed in front of the closets. Silence.

She casually opened the door and there stood Jenkins, smiling. "Room service."

"Mm-mm, looks pipin' hot to me," Latanya replied, smiling in return.

Jenkins stepped in, and Latanya shut the door quickly behind him. While he took off his jacket, she nimbly skipped around him to the desk. She picked up the glass and turned towards him, breaking out an energetic smile; the one she had honed over the years to communicate 'you're the one I want to be with . . . right now,' the one that since January of last year she had reserved for Paul.

He reacted as she hoped he would, with a ten thousand watt smile of his own.

"Before we start, how about a little of my infinity love potion," she purred, offering the glass, "along with a turbo-charger," she finished, holding out the remaining pill.

"Ah think Ah'm good."

"Oh, but you wouldn't want to disappoint me now would you?" Latanya said, fingering one of the buttons on his shirt. "I've got a few hours to kill." She smiled salaciously, cocking her head to the side while looking up at him.

Jenkins pursed his lips and squinted with his right eye.

Latanya took the glass and sipped a tablespoon of it.

"It's safe, in case you—" she gently stroked a line down the front of Jenkins' chest with her index finger, hooking onto his pants, "don't trust me." She pulled him in closer.

"What's not ta trust?" Jenkins replied, taking the glass and downing the contents, leaving the lime at the bottom.

"I'll take care of you--" she said, as she pushed him onto the bed, "and maybe you'll take care of me." She straddled him and started loosening his tie.

"Ah think that can be arranged," he said, playfully.

"Well that's just fine with me," she cooed, "because where I'm sittin' I'm beginnin' to feel some front-li-ness," throwing his shirt open.

Latanya leaned back and undid his shoelaces. He kicked his shoes off.

He sat-up partially, pulled his t-shirt off and quickly proceeded to his belt and pants.

She got up and slowly pulled his pants off.

He took his socks off. His only remaining adornment was the pair of bright white briefs, bulging in front.

"Ooh, now is that some Southern hospitality aimed my way?"

"Sure is." He slid the shorts over, and then below his tumescent organ.

"My, oh my . . . we are gonna have some fun today." She leaned back down to the bed and gently flicked her eyelashes around the bottom of his engorged piece. After half-a-minute, she asked: "You like that?"

"Mm—hmm." He moaned. "Can't complain," he murmured. She moved up higher. "And how do you like that?"

"Mmmm. I'm beginnin' to feel properly taken care of."

"Good," she replied as she sat up and unscrewed the top of the lubricating jelly tube. She then squirted some on her palm. She move her hand up again, then sideways, slowly, in a screw-like motion until she reached the bulging head.

"Hmm, Liz Honey, now that feels better'n pullin' a lever, doesn't it?"

"And how." Latanya smiled, as she repeated the movement.

"Ouch. A little tight there... how about a little kiss to make up for it?"

"Oh, I will," she purred, "but let's just let this work itself in first – it'll increase the pleasure."

She felt him grab her left hand firmly, almost violently. "Ah think my tower of power would like a little kiss." He wasn't smiling so much as glaring at her.

No Saliva. No DNA.

"Well, well honey, I will tend to your tower with my lips," she said with a coy smile, "I've been dreamin' of little else all day now . . . but, if you'll allow—" she said, feigning utmost respect, "I'll apply some lip balm first. I don't think the nerve endings on your mighty oak would appreciate scratchin'."

He loosened his grip and smiled.

She got up slowly, and sauntered elegantly towards the desk.

With her left hand, she took the lip balm out of her purse. She took the cap off clumsily with her forefinger and thumb, letting it drop to the ground. She would normally doing this with her right hand, but using her left hand bought her time, which she needed. *Five minutes for that drug interaction to work.*

She turned towards Jenkins and puckered her mouth. She then applied the balm, a millimeter at a time. But with her left hand, she didn't have good control and the balm stick slipped into her mouth.

"That's the right move, sugar, but the instrument's a little small," Jenkins quipped.

Latanya, lowered her hand and puckered her lips again. "Honey... consider it practice."

Jenkins scratched his throat.

"I'm gettin' thirsty. Bring me some water, will ya Lizzie?"

"Right away, sweetie."

She went into the bathroom. She heard him hit his chest with his arm. *A critical turning point.*

He started looking up. "I don't understand something. Is the . . . room . . . turnin' blue?"

"Yes sweetie, I think it is. Listen, I've got a warm-up act. You'll love her."

There was a long pause. His breaths were short. "Whatever. I'm feelin' good n' horny."

Latanya stepped to the closet and opened the doors. Clarice was still sitting crouched, looking straight ahead at the bottom hems of her dresses.

"Clarice, it's time for the fun. Get up."

Goodson exited the closet.

Latanya led her to the bed. "Now look at that beautiful piece of work," she said, pointing to the erection.

Goodson stared without expression for a few seconds, then nodded.

Jenkins smiled, but with eyes unfocused.

"Clarice, I think he would like his tool kissed."

"Okay," Goodson responded as if she had been told to pay the driver at the front of the bus. Goodson kneeled on the end of the bed, putting one hand on either of his thighs. She slowly lowered her head to the bulging organ.

Jenkins moaned. "It feels great down there."

Latanya stole a glance at her watch.

"But Lizzie, it's odd, I don't feel my left arm." The rate of his breathing picked up even more. He began perspiring.

"Funny, isn't it, how sex just makes you feel tingly all over, right?" Latanya answered. *Myocardial infarction starting, according to Artois.* "By the way, I was sent here by the team," she said as she took the tray holding the deadly back-up concoction and the glass and deposited them in the bathroom sink.

Jenkins nodded weakly. "Can't breathe." He suddenly looked out straight at her. He held his hand against his chest. "I'm not . . . what's happening?" His eyes looked upwards.

Now was the moment, she had to extract information, while he was still coherent. "Do you wanna guess by whom?" Latanya asked as she exited the bathroom. Jenkins was now holding his neck with his right hand.

"I dunno… Antoine?" he said, wheezing.

"No, guess again," she asked, hoping he would divulge some names, names that Paul could connect to faces.

308

"Jeez, I'm feelin' like—" he coughed loudly, "shit... o.k. Ken-" he said, hitting his chest. "Oh man. What the fu—? Kennry ... Silver Fox," he said sitting up, his pupils now circling the room.

"Why him?"

"He—" Jenkins coughed twice. "Can't breathe. Was here," Jenkins eked out in a raspy staccato.

"Wrong, you son-of-a-bitch!"

Jenkins' eyes bulged as he tried focusing on her, his eyes points of intense anger, like neutrons concentrated in fissionable material. He attempted to sit up, at one point swinging his left leg up, but Goodson was pinning his legs down. He still focused his anger at Latanya, through hard, squinty eyes. His complexion appeared gray now.

"Robinson sent me."

Jenkins leaned forward and then fell back.

"So why am I here, what's with the voting machines"

"N-no, bi-bitch."

Latanya approached him, removed her bra, and held her breasts firmly in front of her. She threw her head back, letting her hair radiate, then cocked her head firmly to one side, and began slowly, seductively, caressing her breasts. "If you tell me, I'll let you s-u-uckle them."

A crazed expression crossed his eyes. "Glitch, one ... "

Latanya strained to hear, but couldn't. "Again?" she asked, gently, pushing her breasts up and forward.

"four ... eight!" he pushed out, eyes bulging, his face red. He puckered his lips and clutched the bedsheet, writhing. Suddenly, the twinkle left his eyes. They were lifeless. His brow was still furrowed.

The bed creaked gently as Goodson continued the up and down motion with her head.

Latanya picked up her bra and put it on. "Clarice, you can stop."

Clarice stopped, her head suspended above the stiff organ.

Latanya stepped to the armchair farthest from the bed. She steadied herself using the armrests, then straightened up. She glanced at her hand. It shook like an autumn leaf in a breeze.

309

Again, she breathed in deeply, conscientiously. She reached under the bed and picked up the gloves she had thrown there earlier. She slipped them back on.

She stepped to the desk, picked up the wet hand towel and wiped her lipstick imprint from the rim of the glass. She then picked up the hand towel and the plastic tray from the nightstand and took them to the bathroom, where she washed and dried the plastic tray, replacing the bottles she had removed earlier. She rinsed out the hand towel three times, then washed her rubber gloved hands with soap and water. She took a fresh medium towel and dried her gloved hands.

Latanya then turned on all the lights in the bedroom and stepped to the bed.

Goodson's head was still elevated, staring at the wall painting.

"Good job, Clarice, you've done a good very good work. Now just go to sleep on the floor at the base of the bed, until you both wake up again."

"Okay," Goodson mumbled.

Latanya glanced at her watch. This whole event had taken forty-eight minutes so far. *Now to finish up.* She walked to her purse and removed the vial with the GHB. She wiped it down with a tissue, stepped to the bed, put the vial into Jenkin's open, claw-like right hand and pressed his fingers against it. She picked up the tube of lubricating jelly and did the same. She let the tube drop onto the bed. The vial she picked up. She stepped back to the desk and picked up the remaining 'popper' tablet. She walked to the other side of the bed and placed both the vial and the white tablet in Jenkin's jacket pocket.

Latanya got dressed, put on her shoes and then did a final check. One blue-green pill on the desk, with an emptied drink glass. The bathroom was all neatly arranged with two hand towels. The closet doors closed. The well-sculpted corpse in the middle . . . the young woman curled up like a cat at the foot of the bed. But something was wrong. *Clarice doesn't deserve this.*

Despite her survival instinct screaming for her to clear out quickly, Latanya put down her purse. She opened the closet, and took out Goodson's coat, dress and blouse that she had so neatly folded earlier. She walked to the left side of the bed and turned the skirt zipper towards Jenkins' hand. She pressed the zipper knob against the index finger and then against the thumb. She brushed a little of the fabric adjacent to the zipper against the middle and index fingers.

After that, she took the coat and pressed the top button against the stiffening thumb of Jenkins' right hand. She brushed the inside edge of the left flap against the middle fingers. Finally, she took the blouse and touched the second button against the same fingers.

Latanya stopped, looked up, and scratched her head. She buttoned up the blouse, before taking blouse in her right arm and simulated a ripping motion with her hands. *No, do it like you mean it.* This time she simulated a violent ripping motion. *Good.* She stepped back to the bed, pressed the inside right upper lapel of the blouse against all four fingers of the blonde corpse's left hand, then repeated the same for the inside upper left lapel.

She stepped back between the bed and the door. With a gloved steely grip on either side of the lapel, she ripped the blouse open.

Three buttons flew off, one she heard hitting the inside of the door. The bottom two buttons still held on. *Goddamn good workmanship,* so she ripped again. The buttons finally gave and the blouse was now fully open. Latanya was pleased.

Suddenly, there was a knock at the door.

She froze, both arms up and the blouse stretched wide open. She closed her eyes. *Dear God! Just not housekeeping.*

The moment stretched to a millennium.

"Clarice, are you going to Governor O'Donnelly's talk?" someone asked.

I hope they didn't hear the button ricocheting off the door. Latanya tiptoed to Goodson. "Ignore them. They're enemies," she whispered.

"Clarice, you in there?" the man asked.

"Come on, Henry, maybe she already left," Latanya heard, from someone standing next to the man.

"She could be holding our place in line," a youngish female voice suggested.

More footsteps. Silence.

311

Latanya lowered her aching arms and returned to the spot near the door. She lifted the blouse higher and flung it on the floor.

Latanya stepped back and reviewed the scene. With the torn clothes, the discovery of the GBH in her bloodstream, and her panties still on, Goodson's reputation would remain intact, but Jenkins' – that was another story. Latanya picked up the coat that she had laid on the other side of the bed, and walked towards the door. She dropped the coat near the door, in front of the blouse. Finally, she turned down the temperature on the thermostat. The fan clicked on.

She picked up her purse and walked to the door. She pressed her ear to it. No one. She turned off all the lights, leaving only natural sunlight basking the room through a crack in the curtains, and turned the door handle.

She peered into the hallway. One straggler running down the hall away from the door. No more footsteps.

Latanya flung the door open, stepped out and just as quickly shut it. She walked briskly to the stairwell, entered it and suddenly felt so dizzy she had to grab the handrail to avoid falling. She sat down on the steps. *Can't stay on this floor.* She took the pin out of her hair and let her hair fall. Nauseous, her head spinning, she managed to step slowly down two more floors, all the time holding on to the handrail with both hands. She sat down again just past the third floor.

She recalled her yoga breathing exercises. Deep, calm. She repeated three times before opening her eyes.

The door to the fourth floor opened and she instinctively rose. She retraced three stairs up to the third floor door, all the while hearing the rapid click-clack of shoes above her. She tore off her right glove and then her left glove, making sure they were inside out, and stuffed them into her purse. She entered the third floor hallway, sucked in a deep breath, and nonchalantly walked to the elevator.

From the opposite end of the hall, two people entered the elevator. Latanya felt she would collapse at any moment. She stopped and crouched down, pretending to adjust her shoe strap. She sensed one of the attendees glancing in her direction.

The elevator door closed; they weren't going to wait for her and that suited her just fine. She straightened up, walked to the elevator bank and pressed the button.

Once on the first floor, she again felt faint, so she leaned against a square column.

Ten more stragglers walked briskly towards the hall where Governor O'Donnelly was being introduced.

After two minutes, Latanya joined the dwindling line, now advancing quickly. As the well-groomed people moved forward, she passed a cameraman filming a boy of about nine who was telling the cable channel viewers, "My parents earn the money, why should they . . . they have to pay anything to the government? It's their money." She reached deep inside herself to avoid turning around, sitting the boy down and explaining what a social safety net was. Instead, she smiled widely to better project the image that she was a proud attendee of the Grassroots Political Action Convention's Thursday Keynote address.

She collapsed in one of the few free seats at the near end of this, the largest ballroom of the hotel, and looked out over the enormous gathering of clapping people.

How will the bad guys react when the news hit? It would be by tomorrow morning at the latest, but the GHB could wear off earlier. *Late afternoon?*

Fifty-five minutes later, the tenth and final applause died down and people started streaming from the ballroom. Latanya, somewhat recovered from the fainting spell, followed the part of crowd exiting the building. She continued walking down the sloping street until she reached the Metro entrance. She glanced backwards and saw only three college students behind her. *Safe.*

Once at the Union Station train station, she removed the two gloves from her handbag and dropped them in a garbage can. She walked another fifty yards and held the plastic self-sealing bag with the vial of potassium cyanide over another garbage can. *Maybe not.* She put it back in her bag.

313

Once on the train, Latanya took out her iPad and in the note app wrote two names down: 'Antoine,' 'Kenny Silver Fox'. She also wrote '1-4-8.'

Forty-five minutes later, the train entered a tunnel. Leaning back in the seat, she turned towards the window. The face she saw reflected sent a cold chill down her spine. She looked down at her legs. They were shaking uncontrollably.

Latanya got off in Baltimore and took the subway to her hotel. Her meeting with the Johns Hopkins professor would start in 80 minutes. She had an hour to take off her hair extensions, shower, dress, and collect herself. The meeting would be vital, not just for reasons of alibi should she need one, but more importantly at this point, to keep her mind off of what she had just done.

47

The wind sweeping through Baltimore that evening froze beetles onto puddles. Latanya walked into the Seven-Eleven, and her nose started running. She took a tissue from her purse and held it to her nostrils as she made a beeline for the nut section. There, Latanya pretended to rifle through peanuts, finding none satisfactory.

She failed to find any notes. Her heart started beating wildly. Her thoughts momentarily returned to the day of the fire on North 6th street. *Not again!*

With both hands, she started groping every half-inch of the shelf. Finally, behind a fallen bag of cashews, she found the neatly-folded note.

An hour later, she was sitting on blood-red vinyl booth cushions of a Denney's. Michael sat down across from her, wearing designer glasses, his head shaved bald.

"So Sis, whazzup?"

"One dead Texan skank."

Michael threw himself back into the booth chair, shook his head and smiled. "Wow. You're something crazy. How'd you do it?"

"No can tell. Ever," she said, cool as crypt marble.

"I won't push," he replied, shaking his head from side to side. "So, I suppose you wanna see your 'B'-'F'."

Latanya nodded.

"Well forget about it. They're probably on your tail right now."

"The body was discovered no more than half-an-hour ago . . . and I was very thorough. Gloves all the way. No witnesses."

He leaned back and smirked.

"Michael. I'm clean, I'm sure. Nobody following me, no way," she insisted, sensing that she sounded desperate. "Listen, I've got some names, I've got to ask Paul who they are."

315

Michael furrowed his brow.

She knew he couldn't deny his older sister anything, when push came to shove. He could only justify his actions to her rationally. And in this case, he had no arguments.

"Okay, Tan. I hope he's in one of his 'clear' moods."

Paul was reading when they entered the old three-story townhouse. It was completely unremarkable, in a derelict section of town, which described about forty percent of Bal'more.

Latanya understood why her brother had moved Paul here. On top of the incredible anonymity one could find in a slum a thousand times larger than Harrisburg's, it meant that even if Pennsylvania State Troopers had convinced their Baltimore brethren to help out, the latter wouldn't find them easily in this urban haystack.

Latanya touched his shoulder. Paul turned around. He smiled widely.

"Lat, Lat."

"Yes, it's me."

He smiled. He extended his hand halfway to Latanya. He couldn't seem to move it further.

"Are they treating you well here?"

"Goo', good. But want food, hummus,"

"Sure, honey, I'll make sure Michael brings you some."

"Tanks."

"Paul, I have some names of people who might be doing us bad. You might know who they are."

Paul shrugged his shoulders and moved his pupils from left to right.

"Does the name 'Antoine' mean anything to you?"

He pondered a full minute before shaking his head.

"What about 'Ken or Kenny Silver Fox'?"

"Ken? Kendrey?" Robinson asked. He looked intensively at the wall. "Kendrey good. Computer guy. He hit agent."

"Wait. You say he was good? ... Hmm. Maybe it's a bad guy with the first name of 'Kenny.'"

"Kendrey helped me."

316

"Sure, sure. But 'Kenny' is a popular first name. What about 'Silver Fox'?"

Paul chuckled. "Silver fox. Stupid spy flim name."

"Anyone?"

Yawning, he scratched his head. "No."

"Okay, sweetie, that's it for names."

"Good. Wanna hear joke?"

Latanya smiled. "Sure, whaddya have?"

"Dal-, Dalai Lama go to New York. Hungry—" Paul paused, trying, it seemed to Latanya, to keep the thread. Then his eyes lit up. "Yeah. Go to hot dog stand. And he, he orders hot dog. Wait. He ask –that right – he asks, 'Make me one wi' everything.'" Paul smiled, proudly.

"He's been telling that joke twenty times a day. If I hear it again, gonna finish yo' man off myself," quipped Michael.

Latanya glared at her brother.

"Jest kiddin', Tan, but it would help my mental state if you could whisper a couple of new jokes in his ear, maybe surprise me."

"Michael!"

"Man, Sis, how long can I keep doin' this. I ain't workin'. None of ma two jobs. I'm going to have to start collectin' welfare."

"Michael, I've got these names. I'll take them to the police."

"Yeah, yeah. The baddies gonna have so many people out lookin' when their point man dies, we're gonna be livin' under full court press by the L.A. Lakers."

"L.A. Lakers!" Paul burst out. "Shaque attack. Shaque attack."

"Dude, the Shaq hasn't played with L.A. since o' four," Michael responded, before turning to Latanya. "And that ain't all, Sis. During the day he gets restless, wants to step out on the town."

"Maybe you can do that tomorrow. Just once. It could jar his memory."

"Maybe I'll take him to a Ravens game, too. I'll tell everyone 'I'm escortin' an escaped fugitive around the states and he 'jes' loves football'."

317

Latanya grabbed her head in both hands. "My God, when will this end!" she screamed.

Silence reigned for an eternal ten seconds.

"Ravens, c-crap. I want see Steelers," Paul murmured.

Latanya's laughter was unchained. She guffawed for a full minute. Paul joined in. As her laughter subsided, she realized he was putting whole sentences together. He was improving.

"Michael-" she said softly, "can we talk about this a little later, I want to ask Paul some other things."

Michael shrugged his shoulders and walked away.

She turned to Paul. "Paul, do you remember what happened in Hollidayton?"

"Mur-, Murder."

"Who?"

"Journalist. Guy with an erection...Policeman."

"A police journalist... with an erection?" she couldn't refrain from asking.

"N-n-no," he laughed, before continuing, his brow deeply creased. "Guys in black. Step into café. John run out with computer. Tells me guns not stanart issue gun. John save me."

"Wait, sweetie, I'm going to write this down." She took out the notepad and wrote down what he had uttered, not understanding anything. "Who is John?"

"Boy." Paul pointed to himself. "I gave twenty dollars to use computer."

"What did you use the computer for?"

He winced. Clumsily, he scratched his head. "Send e-mails. To Elaine. To station guys."

"Honey, did you send an SMS, you know a cellphone SMS to Elaine?"

He squinted again. He looked up to the left, then cocked his head to the other side. He looked puzzled.

"Yes. I, I asked to call journalist."

"Was it a lawyer's phone?"

Again he creased his brows in deep concentration. "Maybe. Lawyer we took car from."

"You carjacked him?"

Paul appeared lost in thought, then smiled. "Asshole."

318

"Honey, did you shoot him?"

He pondered again. He nodded. "Wouldn't give car."

Latanya reared her head back. *Stay calm.* "You . . . killed . . . him?"

"No, n-no. Alive. Shoot coat," he said, before grinning again.

Latanya was confused. She bit her fingernail. "Paul sweetie, was he alive when you last saw him?"

He nodded. He grimaced, then slowly bent all fingers down on his right hand, leaving only his middle finger up. He thrust his arm in the air, rabidly, again and again. "He ran after us. Finger, finger, finger."

Latanya couldn't suppress her laughter, emerging like sparks from a freshly-lit roman candle firework. Paul joined her in laughter. She rose up and hugged him. "Oh Paul, I love you so much."

"Me too," he said, pointing at himself, then at her.

She kissed him on the lips.

"Watch out, guys. The nurse just tole me he can't get too excited."

Latanya turned to Michael, fixing him with a steely gaze. "Why don't we let him decide?" She turned back to Paul.

"Paul, are you ready?"

He looked up in her eyes, his whole face radiant, and nodded. "Oh yeah. Missed you."

Latanya's heart beat faster. She wanted so much to be with him. And any day now the evil ones would know he hadn't been burned to a crisp in that fire, and all alerts would be sent out. The villains in this real-life drama would follow her and Paul to the ends of the earth, condemning the two of them to live apart for weeks, for months . . . maybe forever. But not tonight.

"Michael, would you mind walking around the block for a couple of hours?"

"Sis, you cra-azy," he said, throwing on his jacket. He exited, slamming the outside door shut and locking it.

She knew he was ready. How could he not be? It had been a whole three months. She took her shoes off. "Paul, we haven't kissed in a while. Would you like that?"

He smiled. "C-Crazy not to."

319

She smiled and took off her top. She started undoing her bra.
"Can do that!" he called out.

She sat down on the bed. She heard him taking his t-shirt off
and felt him struggling with the bra clasp. She giggled and he
joined her.

"Nervous. Been long," he murmured.

"I know, honey. It's been long for me, too." She took and
kissed his hand. "Let's just take our time."

Paul unclasped the bra, and then knelt down in back of her.
She felt his warm chest on her back. He caressed her cheek.
Slowly. He seemed to relish every inch of her jawbone as his hand
lowered slightly, to her neck. He gently cupped her breasts.

Inside, she was quivering. She so longed for this touch. All her
troubles, evil deeds fell to the ground; she was drifting away on a
cloud.

He brought his head forward and she met his lips with hers.
Ever so gently, he caressed her breasts. A combined sensation of
warmth and electricity cascaded through her entire body.

He enveloped her with his arms and brought her in. They
kissed deeply. Her stress melted away as she closed her eyes. She
reached down and unbuttoned, then unzipped his jeans. He lay
down, and she pulled them off.

He hooked the edge of her panties with his thumbs and while
kissing her, slid them off.

She so wanted to feel him.

They embraced. He gently pushed her forward with his chest.

Latanya resisted. "Maybe we should try it the other way. Less
stress."

Paul nodded, then lay down. She straddled him, as he lifted his
head up to kiss. She planted her lips on his, her tongue touching
the end of his. She found and gently put him inside her. She let
out a gasp.

He embraced her arching back and cupped her left breast. She
left him room to breathe, as she fell and rose, ever so slowly, every
few seconds.

It was over soon, too soon.

"Too excited. Too excited," Paul repeated.

"Don't worry, sweetie. It's okay."

320

He seemed dazed. *Oh-oh. Something popped in his head.* He seemed to sense the concern, and kissed her shoulder. "I'm fine . . . hungry for more." He raised his eyebrows three times, like Groucho Marx.

She couldn't suppress a chuckle. "But maybe you want a glass of water, first?"

He leaned back, nodding. "You read minds."

She got up and poured a glass of water. She reflected on her younger brother's patience. Bringing water and all the supplies, for what was now three solid months. An immense debt of gratitude she owed him.

She brought the water and Paul chugged it down, looking all the time in her eyes.

"Now ready, beaut-ful," he purred.

This time, it was slow. She guided him, slowing him down when she sensed he was getting over-excited. They came together.

Afterwards, lying spent next to him, she felt an urge to confess, and he was receptive, as he'd almost never been before. This whole experience had somehow transformed him. It wasn't his words, now awkward; it wasn't the manners or his humor. She understood the transformation as only a woman could, on a level no college professor could explain. "Paul, we're a unit now."

He smiled and cupped her hand in his. He pressed his warm lips against the tender veins of the back of her hand, then looked up and smiled at her.

"I've done the most extreme thing one can do to protect you," she confessed.

"What?"

"I'll tell you next time."

He jerked his head back. He nodded in acknowledgement. "Let's ge' married."

She smiled. "Paul, before you say that . . . get well. Weigh the decision," she said, while at the same time feeling warm inside. Secretly, ecstatic.

"I have," he replied, looking deep into her eyes.

48

Two days later, the press revealed the devastation that she had anticipated they would. Her fear of that moment had been increasing with every passing hour since she left Baltimore. *A double-barreled shotgun blast into the buttocks of some evil organization; not enough to kill it, but enough to make it limp for a little while, and in the end, make it angrier.*

'Police are investigating the death of a man in a hotel room of a hotel in Northwest Washington, D.C. The man, a U.S. army sniper and ex-military contractor was found dead in a room at the Merrio Convention Hotel. He appears to have died of a heart attack, according to Washington D.C. Detective William Seiffert.'

At the same time, in the Patriot News, the article read – 'Identities of the victims of fire on North 6th street now clear.' Six names figured on the list, and Robinson's was not among them.

That afternoon, she got a call on her cellphone.

"Hello, Miss Forrester?"

"Yes."

"It's Olson. I'd like to ask you a few questions."

"Sure, but I don't have any new information."

"Yeah, but we do . . . and we think you might be able to comment on it."

***

The escorting officer passed the aluminum-framed reception area, then walked past desks of men busy poring through papers, photos and even a phonebook. *In this day and age.* As she followed him down the hall, she wondered what information the police had.

As she entered a spartan room, lit with a bare fluorescent lamp, Officer Finnegan greeted her with a nod. Next to him sat Olson, looking glumly at the steam coming from the styrofoam cup.

She sat down on the gray metal chair, itself a victim of a multitude of keys and sharp scratching objects.

Olson leaned back, fixing his gaze on her. "So, you think you're real smart, Miss Forrester?"

Latanya cocked her head and smiled. "Whatever would you mea--"

"Miss Forrester . . ." Olson interrupted, picking up his coffee, which, judging from the aroma, could only have been the barely tolerable public service variety. He took a sip. "You are guilty of a serious felony."

A chill descended her spine like a slinky going down the stairs. *Does he know?* "What?!"

"Don't play stupid, Miss Forrester, please. You could have had such a promising future."

Latanya felt the blood leaving her fingers. *They know.* Was it the cyanide missing at the jewelers? Did someone track her in Baltimore? In Washington? They must have found the fake driver's license. *Oh, no, they followed up with the hotel bellhop.*

"Miss Forrester, we know now you arranged the harboring of Robinson – let's just call him Paul—on 6th street, and now your brother is giving him cover."

He paused, taking another sip of coffee from his cup, this time protracting the sip to an awkward length. He looked in her eyes again.

Latanya looked away and shrugged her shoulders. Relief. The blood rushed back to her extremities, her body temperature dropped to normal. "Oh, *that* . . ." she felt like replying. Instead, she looked down, trying to portray shame. "Lieutenant, I still love him; I hope he is still alive and I promise to help you with everything I—"

"Cut the act!" Olson bellowed. "You made believe you weren't interested in that 'pedophile' anymore, but the gig is up. Your brother is going to Philly for more than just drugs."

She raised her head defiantly but said nothing. She dropped her head back down.

"So, where is he now?!"

She pursed her lips, and looked Olson piercingly in his left eye. "Lieutenant, I can't tell you that, because I don't know. I don't know, because, well, the person hiding Paul—"

"Your brother, Miss Forrester, Michael."

She looked up, finding it hard to focus on them. "Yes, my brother Michael . . . wouldn't tell me."

Olson shook his head and chuckled. Then his smile faded and his eyes flashed brightly. He hit the table with the fat palm of his hand and it shook violently.

"Do you think we're stupid, Forrester?"

"Do you think you're the only ones pursuing Paul?!" she fired back.

Olson raised his eyebrows. He lifted his frame up from the table. "Spill it."

"I saw Paul five times when he was hiding on 6th street, and he told me things."

"And you hid this from us? Look, Forrester, you're not stupid. You do realize you were harboring a fugitive. Five years in prison that'll get you."

"Officer, the first thing I had to figure out was if the love of my life was a pedophile killer maniac!" Latanya blurted out. She looked away wistfully, to some graffiti on the wall. "I also wanted closure."

Olson shook his head. "Look, I'm going to have to book you for obstruction of justice."

Latanya shrugged her shoulders. "You know, Lieutenant, go ahead, if you have to. I can't help you on that point, but I can tell you what else Paul told me—"

Olson cocked his head. He clicked his pen by hitting the plunger against the table. "Agent Finnegan here, whom you met last time – he can smell bullshit from a county away . . . so, this better be good and re-le-vant," he said slowly, as if he were talking to a mentally-challenged person.

Latanya took a deep breath. "By the time the lawyer had complained, you know, a couple of days after the events on the fourth, I wasn't paying attention anymore. When you hear that the most important person in your life killed three people because he

324

was defending his pedophile ring, you might be forgiven for letting slip some comment about a car-jacking."

"Go on," Olson said, spinning his finger in a circle.

"Anyway. Paul mentioned, between a lot of gibberish, that he had taken a car from a lawyer. And his last words on the subject were: 'lawyer pissed. Gave finger, finger.' Then Paul bent the two side fingers down and smiled." Latanya repeated the gesture.

"You mean he left the lawyer alive?"

"Yes. And that's not all. Paul has a friend. A doctor who does some work for WQHP on medical issues. She remembered receiving an SMS from a lawyer. She deleted it, she told me, because it was unintelligible, and coming from an unknown lawyer's phone. But it mentioned that Stoltz-something-or-other had been murdered."

"Stoltz? That could mean Stoltzfuss." He turned to Agnelli. "That confirms it, Joe. He was there."

"Yes. But if he wrote 'was murdered,' then we know he didn't do it, right—" Latanya added.

Olson nodded. "Not necessarily. But let's say he knows who did."

"And she told me something else."

"What?" Olson asked, tapping the table nervously with his fingers.

"The doctor's boyfriend had called Paul that morning, and told him that a journalist had interviewed someone who had claimed the machine changed his vote. It was in the Altoona Bend. And the boyfriend remembered that Paul was very excited about that information. Very pleased."

"That changes things. Rayman was a red herring." Finnegan contemplated aloud. He stared as well at the bottles on the wall. "But we won't be able to confirm his shoeprints until the Spring."

Olson scratched his chin. "Do you remember what shoes he wore?"

"Yes, he wore Clarks Pulverize when he wasn't on camera. We bought them together.

"Size?"

"Ten."

"We'll try to match up the footprints. But it'll be tough on account of the snow cover. It thawed and snowed again since then. Did he remember anything else?" Olson queried.

She looked behind the captain and shook her head. "He was spouting a lot of gibberish. It might come back to me. Sometimes he seemed to repeat things. The first sentence gobbledy-gook and the second one almost normal."

"What do you know about the people you say are after him?"

"Black Explorer sedan. One guy is tall and has a big scar on his face," she said, tracing the location of the scar on her face. "The other guy is short, with a small scar above his eye, and these big hands."

"Black Explorer?" He turned to Agnelli. The treads we found near the doghouse were from the Explorer, not the Toyota or Chevy." Olson once again stared at Latanya. "Anything else?"

She almost mentioned Jenkins, but caught herself. She felt she had tempted fate twice already, once with the kidnapping and another with the 'execution'. The latter had been an act of self-defense of her future family, a move borne of desperation, not a psychopath's afternoon distraction. *But that wouldn't matter to them.*

"No."

Olson nodded. "Okay, Latanya, now here's the key question. Where are they now?"

"I have no idea."

Olson snickered. "You mean you know, but won't tell me."

"No. Like I just told you, I honestly don't have a clue," Latanya countered. "I mean, bring on the lie detector, waterboard me."

His face was turning red, his hands fidgeting.

"You do realize, Inspector Olson," she began, calmly, "my brother did that to protect themselves . . . and me. Those men are ruthless—"

Olson crushed the empty cup so completely, drops of coffee fell on the table "We'll need more'n what you gave us . . . but for now it'll do. Just remember, he's still a suspect, and if he's anywhere in this region . . . sooner or later we'll find him."."

"I hope so, Detective Olson . . . I hope so and fas-- "

"I'll let you off this one last time, Forrester. But stay in town."

326

"My teach—"

"In town!"

***

When she left, escorted by a tall, stern officer, Finnegan got up and closed the door. He turned to Olson. "Thanks chief." He then leaned foreward, putting his hands on the table. "Guys, I'll stick to her like flies to shit, because five'll get you ten, she'll head to Philly again in a few days, and then . . . we'll nail 'em."

Olson nodded. "Worth a try, Finnegan." . . . *but I'm not so damn sure.*

49

Nurse Burns stayed in Baltimore almost until the end of February, and then, as agreed, returned to Harrisburg. To be safe, Michael moved Paul five houses down and across the street. Getting into a boarded-up house was illegal, but not hard. He wasn't worried about getting busted, since the owners of the houses were usually fancy-assed lawyers who wouldn't be caught dead in this part of town. Investing, yes, but showing up . . . not a chance. Just like the ones he had started to hand-deliver his goods to once again.

Paul's schedule had adjusted to the time before the invention of electricity. Asleep at 8:00 at night and up at 4:00 in the morning. He also napped some during the day. The portable gas heaters kept the temperature livable, but barely. He had overheard Nurse Burns, remarking to Michael, when he was pretending to sleep once, that he was progressing remarkably well now and should return to the real world as soon as he can. She departed for good shortly thereafter.

Michael had hired a woman from the area, a Haitian cousin of a salon owner, to bring food to him. On the advice of the nurse, Michael had also found and installed an old stationary bicycle, as well as procuring a few balls and weights. Paul now spent two to three hours a day working out. Before showering, he heated water over the Coleman stove. He could soap and clean himself to the extent a half-gallon of water would suffice.

***

The worst thing for Paul now was the waiting, especially since he didn't quite know what he was waiting for. The four-month anniversary of the November events would be in less than two weeks. He had no access to the Internet, and the newspapers Michael brought weren't much help in keeping him from going stir-crazy. He bugged Michael for a pistol. Michael finally relented and brought him a 1992 HK USP. "And it's with a silencer, so we ain't gonna sound like a shootin' gallery in here."

After that, he assiduously filled the walls with lead. The old plates were shot up within a day, the old cereal boxes within two. He emulated every shooting position he remembered seeing in James Bond movies and worked until he could hit a half-dollar piece from twenty-five feet while shooting from the hip.

Michael would leave for longer and longer periods, sometimes stretching to a whole day. He arrived after such a break at the beginning of March with news that Latanya would be visiting the following week.

Paul sensed his depression leaving. 'Tan' would bring news on the status of the case his lawyer was building, and female companionship, the two things he needed most, and not necessarily in that order.

"Good news. I'm getting ready to go."

"Yeah, dude, we be more'n' ready to check you out soon. I've had enough o' this camaraderie shit. See yo' ugly face one mo' week, gonna do a 'Hannibal Lecter prison guard surgery' on it."

He laughed. "Start practicing on rats, then. His was fine work indeed."

***

Antoine had now been outside the nurse's house every morning, at different times, for the past week. The waiting in the morning brought out the arthritis in his left shoulder. And every one of those mornings, he cursed the bullet he had taken that made it that way.

He shouldn't even be here, what with his mug in every police station in the state. But Jenkins had partied himself to death *before finishing the job . . . second-rater.* So that left Antoine to finish the 'assignment.'

Of course, he and Parkett couldn't dare show themselves in Pennsylvania before the end of January, even though he had been back in fighting form as of the middle of that month, his arthritis notwithstanding. It was common knowledge in his circles, though that time dulls the intensity of the chase for anyone – as long as he, Girelli and Parkett kept their noses clean, the cops would no longer be on the look-out for them.

He reckoned that with a few minor disguises, he was absolutely safe. Parkett, less so, on account of his bulk, but he'd just call on him to assist on the few occasions when he needed the sheer intimidation Jameson couldn't provide. The rest of the time, by mutual agreement, Parkett'd be comfortably parked out-of-state.

Antoine had worked hard over the years to cultivate the reputation of assuring rock-solid payment and well-thought-out, ultimately successful jobs. But the Robinson assignment had sorely tested the limits of those two parameters and, ultimately, his status. He wanted to reverse that now, and fast.

The nurse exited her house, dressed in civilian garb. *Paydirt.* After she had walked half a block, Antoine had the driver turn the car around and follow her. They stopped a few yards in front of where she was walking.

Parkett opened the rear door, and jumped out. He opened his coat, flashed his gun and told her to get in.

She looked around. No one was near. She got in the car.

"Where is he?!" Antoine boomed at her.

"Who you talkin' about?"

"Don't play stupid. The chimp reporter."

"Don't know who you mean."

Antoine took his gun out and pointed it at her head. "Nurse Betty, ya got five seconds to tell us where the reporter is. After that, I shoot you and then I shoot ya' daughter and ya' two lovely grandchildren."

Her eyes opened wide. "No, no, no! Mista', mista', I don't know who you are, but I'll tell you, don't gotta aks me twice." She looked up. "Lord, please forgive me for what I do. Please unnastand." She lowered her gaze to Antoine. "Two-two-three-three-six Etting Street, Baltimore."

"Baltimo'?"

"That's what I said—Baltimore."

Parkett took out a notepad and wrote the address down.

"Who's with him now?"

"I don't know whatchu talkin' about. He's by hisself."

"Look, Mrs. Brown, we aren't fooling around here, ya' family's life depends on what you say now. Who is helping them?"

She bit her lower lip and looked down. "A man named Sammy. That's all I know. He told me not to ask any more. He's from here. He paid me in cash."

"Sammy . . . describe him to us."

"He's black like me. About six feet tall. Medium build. No football player, though. Got a small scar on his forehead . . . that's all."

"Same as all of 'em. Shit!" Antoine barked. "Here's my number." He handed her a card. "Call me now from your cellphone."

She took out her cell and called him. It rang.

"End the call . . . Now, first thing you do is forget calling from your home phone. If you call Sam from your home phone, I'm gonna have to hurt you and your family."

"No, no. Please don't."

"'Den you'll do like I say. Now get out. You never saw us."

***

It was 'serious,' Elaine had declared . . . and 'confidential.' Latanya had subsequently driven down to Harrisburg early on Friday to meet with her in the Bookstore Cafe at one o'clock. This time, she had borrowed another colleague's car, once traveling incognito.

Elaine looked upbeat, as she greeted Latanya. Almost on cue, the two made a show of exchanging pleasantries. They ordered their coffees and took their usual spots upstairs in the bookstore.

Latanya assumed a deadpan expression. "So Elaine, what's the word?"

"The day of the Hollidayton shooting, there were some strange goings-on at the office."

"Yeah, didn't you mention there was some disease outbreak at the cafeteria?"

"Yes, exactly, and I followed up on it. I specifically followed up on the deliveries of vegetables and eggs to the cafeteria that

331

morning and the previous day. In short, anything that anyone would have eaten that day."

"Yes?"

"Every single one of the suppliers had done a thorough check on their produce. I won't bore you with the details, but the Salmonella turned out to be the Salmonella Enteridis form. The suppliers all came out clean." Elaine leaned back.

"I'm sorry, Elaine, I zoned out in high school bio. What would that mean?"

"Oh, pardon me, I've been talking too much with food service people. If it were Salmonella Typhimurium, it would most probably mean that it was contaminated somewhere along the stage of fresh processing or delivery. However, this was a pathogen introduced by someone after delivery, most likely from frozen chickens, which the cafeteria kitchen doesn't use."

"Huh?"

"It means that someone came and spread the live virus in the kitchen."

"Salmonella? Couldn't someone have died?"

"One victim was in the hospital for three weeks," Elaine replied, nodding.

"Wow."

"And we isolated the DNA of this particular virus, to locate it."

"And?"

"It came from a lab at Penn State."

"How did they give it, I mean isn't that a high biological risk item?"

"Exactly."

"Someone called on the lab, said he was from an environmental group and they were told there were some 'dangerous viruses' that could be spread from big industry and overcrowding. And guess how the lab tech described him?"

Latanya shrugged her shoulders.

"Funny, blonde, cute and Southern."

She tried to swallow, but couldn't. Silently still, the cup burning her lips, she processed the statement.

332

Elaine nodded. "Apparently, he charmed the lab tech into showing him around. Showed him the TB, Salmonella, Hepatitis isolates in the refrigerators. He must have stolen a couple of vials, according to her."

"Can someone just do that, just go into a lab like that?"

"No. She was reprimanded, of course."

"But how'd they get the viruses to the kitchen."

"Several suspects are being questioned, but the strangest is this one temporary worker, an admin assistant came into work that afternoon to—" Elaine showed a quotation mark sign, "'replace' one of our admin assistants. There had been no official request. The Station Manager let some cute brunette up onto the floor. Mark, the Producer, thought one of the associates called her in. He was preoccupied at the time, he said, and didn't pay too much attention."

"And her role in this?"

"I don't know, but he said she was circulating among the cubicles for quite a long time. He thought she was passing on some feeds and wanted to make sure the people got them. In fact, no one had given her anything to pass on."

"No."

"Uh-huh. And he said she went to Paul's desk and later someone saw her coming from my cubicle."

"What?"

"You know, no one shuts their computers down, right?"

"Yeah."

"Well, Paul's not stupid. He would have texted, e-mailed, S-M-S'd something to us. I thought he had had sent an e-mail with the subject line 'Murder', but you know how he was always joking."

"Uh-huh . . . and he wouldn't have sent it to everyone."

"Exactly. I think she went into our e-mails and deleted his messages. My password should've been tougher, but I never thought--"

"Birthday?"

Elaine nodded, with resignation. "Enough to warrant paranoia, isn't it?" Elaine concluded.

Deep in thought, Latanya nodded.

They left the cafe, and once in the street, Latanya turned to Elaine. "There is one more thing. Paul said it was extremely important. Could you please take these car keys, pick up the burgundy red Chevy with this license plate at this parking lot—" she said, handing a post-it note to Elaine, "And drive to this address in Cumberland. The man living there is Marty Leboeuf, with his wife and two young children."

"And, why, may I ask?"

"I can't explain why, but it's linked to the voting machines. Please ask him precisely this: 'Is it just the U.S. Senator?' I can't explain what for, Paul wouldn't say."

"Is it just the U.S. Senator?"

"Yes, just that one question. But look for his reaction. Paul said you would know right away if he pretended he didn't understand, or lied."

"Latanya, I don't—"

"Paul said he would never put you in harm's way, but we must know and I've already probably been tracked to testing the machines . . . it sounds crazy, but please, trust Paul on this."

Elaine nodded slowly. They agreed that Elaine would accept phone calls from unrecognizable numbers, as Latanya would be calling from payphones and the cellphones of strangers.

50

Latanya checked that the brushes, the coal powder, the ultraviolet lamp and two invisible ink-revealing liquids were neatly arranged in the briefcase. If anything had been marked or written on the outside of the voting machines, with these tools she would find out. She crossed the street slowly. The gray clouds felt particularly oppressive today – about to drench her and everyone else approaching the tall, authoritarian buildings.

Perrino had brought the machine out and laid it on the floor. Latanya first inspected the cover. There were the usual nicks and scratches, but nothing that set off alarms. She asked that another Danabold machine be brought in, so she could compare the scratches.

Perrino obliged.

There were no differences or distinctive placement of the scratches. The machine was also free of any markings.

Forrester took out the coal powder first, sprinkling it over one side of the machine, then the other. Dozens of fingerprints stood out, as well as some long lines, along with one that started as a thumb print, but morphed, over three inches, into a fat line. Forrester stared at it for a few seconds. She took about a quarter inch from the bottom border of it with her little finger. She put the finger to her nose and smelled it. "Did they have mayonnaise?"

"Yeah, could be. At lunchtime, we served a sandwich buffet," responded Perrino, "They're volunteers after all, and especially for the college kids, it's better than pay."

Latanya nodded. She asked Perrino to shut off the light for a few seconds. Seconds later, the room turned cavernously dark. Perrino's department didn't merit an office with a window, and now this was a blessing.

There were no fluorescent markings on the machine.

Latanya reached into her bag and felt for the wand-like UV lamp. She took it out of the bag. She switched it on and the purplish glow seemed other-worldly. She waved it over the cover from a foot away. Nothing but the lightest brightening of one side, as if that side had a slight deformity. She moved the lamp eight inches closer and started passing it slowly over the machine.

Some figure or number suddenly flashed brightly. She stopped and passed the light over the figures in the other direction. She felt her heart beat twice as fast as it had, ten seconds earlier.

The number '89' was clearly written, on the right side of the machine. *Okay, okay, girl, don't be jumpin' to no conclusions.* She drew in a deep breath and waved the ultraviolet lamp once again over the figure.

"That's odd," Perrino piped up.

"Hmm. But we can't infer anything from this. Not yet. I have to check the other machine." *Can I trust her?* She took a deep breath to calm down. "After all, this may have been some, I dunno, crude internal method for marking batches of machines. The next machine might have a 54 or 105 or something else on it."

"Yes, I see," Perrino responded.

Latanya took the cover off the second Danabold voting machine and swept both sides with her ultraviolet wand. Nothing. The sides weren't marked.

"Anything?"

"No, nothing," Forrester replied. "So I guess we can bag that theory."

Within an hour, they had accounted for seven parameters and excluded six: Votes for any particular candidate—109 votes was the lowest number for any. It wasn't votes on the referendum, as 231 were cast against, and 214 for; write-in votes, only 21 for all senators, congressmen and councilmen.

Latanya concluded that it might be a percentage. She started with that referendum, which would have given gas developers the right to buy land on an otherwise pristine, protected lake. It was a big money project, ultimately hundreds of millions for the people divvying that up. The votes for—43%, against—56%, the rest abstentions. *That leaves the politicians.*

336

"Would you like to start with the Congressmen, while I do the Senators?"

"Okay," Perrino replied.

Twenty minutes later, Latanya felt her head spin. She rose suddenly, pushing the chair back loudly.

Perrino looked over. "Did you get something?"

"Oh my Go--" Latanya started before stopping herself. *I can't let any of my students get involved.* Latanya leaned against the table. "According to my calculations, 400 votes for Dooley, and 49 for Duquesne. One abstention. Eighty-nine percent for the winner."

"Goodness, that's a large difference. Someone really wanted to make a point!" Perrino said, matter-of-factly.

Latanya smiled. *Situation 'de-dramafied.'* "Sure did. How did the Congressmen turn out?"

"So far, so normal. I've got one whited-out sheet and the count so far is about... fifty-fifty.

It is Senator Dooley for sure. But to remove all vestiges of doubt, Latanya would need to vote on the machine herself.

"Good, Miss Perrino, so it's just the Senators that stand out a bit. Do you think I could carry out a vote on this machine? You know, just to remove all doubt."

"Well I suppose so . . . but I'd have to mention that to the company."

"Do you? I mean isn't it *your* responsibility to test the machines' integrity?"

"Yes," Perrino replied, furrowing her brow, "it is, it is," she said, looking up. "On Monday, okay."

Latanya so hoped, as she walked to the car, that several political offices were affected. That all the votes were off, that there was a generalized technical issue with the machine.

***

Antoine called through to his network in Baltimore. No one lived at 22036 Etting Street, although they had found proof of squatters having lived there recently; the nurse hadn't lied. Over the next two days he got nothing more. It was a big area, with

337

heaps of medium-built black men exiting and entering run-down buildings.

The drug dealers cast a wide net, but they hadn't noticed any out-of-place homeys in the neighborhood—certainly none harboring white folk. Not that there weren't white boys in the area, it's just that most of them crashed after taking their doses and then got up and left, or died from overdoses and were thrown in some dumpster or empty lot. There were very few camping out in the hood long-term.

As he received the last info, he froze. *Weird that a completely-unrelated piece of info sparks that light bulb.* They'd been conned. Jameson had fallen for the tears, the church goodbye. The chief had telecons "confirming" they had broken up. *It was all crap.* Forrester's brother hadn't been in his spot but once or twice in the past month.

"Shit!" Antoine muttered, his hand still on the door handle. "I'll kill that smartass black bitch myself."

He found her address in his notes and drove to her apartment. He checked the Taser in his left pocket as he steered the car to the curve. As he opened his car door, he noticed a scruffy man in a beat-up leather jacket turning around, bending down and talking . . . to himself. Antoine shut the door, keeping the guy in his range.

The man glanced at Antoine.

Antoine stopped in front of the apartment building for a second, but kept looking down the street. He was sure no casual observer would recognize him now, with the baseball cap and librarian glasses instead of the fedora and sunglasses he had been identified wearing in Hollidayston, but Scruffy Jacket across the street wasn't a casual observer.

Scruffy Jacket was watching him intently and purposely, despite the 'Street Crazy' act.

Antoine walked past the apartment building to a liquor store thirty yards farther down the road.

As he opened the door a bell rang. It triggered a thought— they didn't need to talk to her. Besides, she might bullshit his crew again, *that wise-cunt.* In the meantime, he didn't want to get on the police radar by chasing her here. And though the Law were slow, this snitch across the street was proof they weren't complete

338

morons. He snatched up a bottle of McKenzie Rye Whiskey and approached the cash register. He peered out the dusty window.

Scruffy Jacket was glancing at the store.

Antoine fingered his new Beretta. He pulled out his wallet and retrieved a fifty-dollar bill. *Discrezione*, his grandpa often whispered to him, when he did something rash or violent. "You'll go further not doing stupid things because of temper," Gramps would add.

Antoine strolled to the car, his right hand firmly wrapped around the bottle, avoiding the cop's gaze all the while.

***

Latanya felt she needed to see Paul. She missed his smile, his soothing voice that he had regained, almost miraculously.

The meeting she had arranged with the Sociology department at Johns Hopkins was supposed to be in two weeks, so she called and moved it forward by eight days. She informed her dissertation advisor about the trip, and no one else. Latanya twirled the end of one lock of her now returned-to-normal, slightly curly hair. She would leave in three days' time, again by train, via Philly.

The cellphone range. It was David.

"Someone called, asking for you."

"What? Who?"

"Didn't leave a name. I told him you were out researching one of our projects. I told him you'd be back on Friday, but the guy didn't want a meeting."

"A guy? Was he from the State Department?"

"No, or at least he didn't say he was."

"Thanks Dave, this. . . may be key."

As she hung up, she continued the thought that had started to take shape. The machines were the Danabold 1-0-4-8 model. Jenkins last utterings were 'one'- 'four- 'eight'. Had she misinterpreted the 'oh' as an 'ugh'? Latanya lifted up her right ring finger for a brief chew. *One more thing to do.*

51

The screen lit up with a welcome to Danabold's voting machine. Some marketing blurb about "cutting edge" then appeared.

"Where's the assistant you were going to bring in?" Perrino asked.

"Tied up with another project," Latanya lied. There was something eerie about this whole situation, and Latanya wanted to avoid drawing anyone else in on it. *Who knows where this may lead?*

She took the ballot forms she had filled out the previous week and put them in front of her.

"Remember, the card is erased and ready for re-use after you pass it through this reader," Perrino said, as the machine's internal drive powered up and the screen lit up with a "Your Danabold 'one-zero-four-eight' is now ready for polling!"

"Thanks. Wait, did it just say one, zero, four, eight?"

"Yes, that's the model. I'll leave you with it now. I'll be sitting over there," she said, pointing to a table with a stack of papers on it, all arranged in neat stacks, "so just call me over if you have any questions."

"Thanks, thanks Mizz Perrino. I think I'll be okay." *This is it. Now to find it precisely.*

Perrino left, and Latanya started on the first of her ballots. Senator Duquesne. She quickly went down the list of the remaining candidates and just marked the top box. These would be the control. If the machine mixed them up as well, then it could very well be that the machine was generally flawed and the 89% a fluke. In her heart-of-hearts, Latanya wanted this to be the case, because the alternative was too horrific to consider.

After fifteen minutes, she had completed twenty ballots. Twenty-five minutes later, another thirty-two. She carefully voted for Senator Dooley, despite the fact that she would soon have to take a potty break. *Only 47 more to go.*

During the last ten, Latanya had to keep her hand from shaking. She decided to change her vote. Purposely, she entered Duquesne once when the ballot was for Dooley. After completing the entire vote, she pushed the summary page. She pushed the 'Y' that followed the prompt asking whether she wished to go back and change a vote. It wasn't particularly warm in the room, but Latanya could feel a drop of sweat dripping down her chin, as she changed the vote from Duquesne to Dooley. Nothing abnormal occurred then. Her summary page now showed Dooley to be the selected candidate.

She finally voted the last one, a Dooley vote, and put the ballot down. There were no 'changed votes.' Nervously, she called out to Perrino.

Perrino came over, smiling. "So, how did it feel?"

"Awesome, to tell you the truth. It's so convenient and, well, Twenty-First century."

"So, I suppose you want to see the total?"

"Sure do."

Perrino pushed some keys and went through a number of drop-down prompts before the 'Voting Ended. Total Now' prompt emerged.

She watched as the totals came up. For some reason, her eye had wandered to the middle of the screen, so the first thing she saw was that the Councilman Nielmann had garnered 100% of the vote. *Well, that part worked.* Latanya then scanned quickly to the top. Her jaw dropped. She squinted, to make sure she hadn't misread what she saw. There, in clear print, the total for Duquesne stood at 10% while for Dooley it was 89%. Write-ins, 1%.

"Is that what your ballots read?"

Her mind frozen, Latanya glanced at Perrino. A lump appeared in her throat. *Can I trust her?* "Not . . . completely."

Perrino smiled and cocked her head to the side.

Latanya, once again plumbed the depths of her belief system, and finally convinced herself that she *must* have faith in the civic process to sort these issues out. If, in the end, she couldn't trust Perrino, then she, soon-to-be professor of Sociology, would have to move not to another state university . . . but to another country. *No!*

341

"Mizz Perrino, the machine changed thirty-nine votes from Duquesne to Dooley."

Perrino's smile vanished. "Are you. . . you completely sure?" she asked, pointing to the ballots.

Latanya nodded, keeping eye contact with Perrino.

Perrino brought her hand to her mouth. "I—how? By how much?"

"I had split the votes evenly in my ballots. Fifty for Duquesne, fifty for Dooley. Wait, I changed one to Duquesne. Okay, fifty-one to forty-nine. That's forty votes flipped."

Perrino shook her head. "How did the other votes turn out?"

"The Congressmen votes I also split down the middle and that's what the machine registered. I voted for only candidate or issue for every other spot. And that was one hundred percent accurate."

Perrino slowly nodded. "Maybe they'll sort it out."

"Who?"

"Danabold."

Inadvertently, Latanya brought her ring finger to her mouth and gave the fingernail three quick nibbles before bringing her hand down again. "But they might be responsible—"

"I just called them, fifteen minutes ago."

Latanya felt the blood drain from her face. Her heartbeat accelerated. "What? Why?"

Perrino lifted her shoulders. "A courtesy call. Their rep asked me to call them whenever anyone takes a special interest in their machines, or we do a parallel test. I didn't want to bother anyone just because you had shown some interest, but a parallel test is another story."

Latanya clenched her hands into fists, digging her fingernails into the skin of her palms. She strained to prevent a torrent of profanities. "Oh, okay," she replied, before forcing herself to smile. "I'm sure it's just some little defect in this machine. Dust or something."

Perrino smiled back. "We hope so, don't we?" she said, rolling her eyes.

Latanya chuckled. "Sure. On another note, I understand why you had the fourth person at each table," she turned back to Perrino, "my bladder is about to burst."

Perrino smiled and pointed parallel to the wall. "Down the hall and to the left."

After using the restroom, Latanya glanced down the hallway in both directions and walked to the exit farthest from the testing room.

She got into the car, tossed the mini-wallet with the fake driver's license and accompanying business card onto the passenger's seat and took a deep breath. *Oh my God.*

52

It was the third car of the 642 from Harrisburg to Philadelphia that she entered at 7:30 in the morning, joined by a crush of people. Seating was at a premium, but a few expressionless government workers from Pittsburgh had just exited the train, making space for her. She threw her backpack on a window seat facing forward. She hadn't noticed the short man with green eyes passing her and taking a seat in the next car, a seat that faced hers.

Upon exiting in Philly, she checked the boards. Her connection would be on track two. Tons of people were milling about on the platform. It seemed to her that, from across the platform, one tall police officer was observing her every move.

As Latanya got on the Amtrak 95 Northeast Regional train, she glanced backwards. The policeman said something into his walkie-talkie. *They're following.*

But shortly after sitting down her panic dissipated. According to McKinley, the state troopers don't go into other states to follow a criminal, and the police doubly so. They might not even leave the city limits. Besides, she was in Maryland for dissertation work and she had written as much in the schedule she had handed to Olson. So she was a few days early. Unless officer Agnelli was willing to take this across state lines, she was safe. *But probably not for long.*

Damn it, if they can't resolve it, if Paul can't tell her anything more, then he would be arrested. It wasn't the jail that bothered her – a month in the slammer until the facts came to light wouldn't bother her. *But what if another 'Jenkins' were to show up on the scene, blowing up a jail cell block?*

After her morning meetings at the University, Latanya's thoughts once again returned to the policeman's gaze. The call. She broke out in a cold sweat as she walked to the station. *There's no way they could have arranged to follow me . . . or is there?*

Before entering the metro, she looked to her left. What does an undercover cop look like anyway? Would it be like the beefy guy with a designer rip-off Lauren coat? Would it be the thin tall one there, reading the paper but with a satchel over his shoulder? Was it that short guy who kept his gloves on, the one wearing a hood over his head? Addict. Maybe the cop is white. There were four white people in this car. The hooded junkie she wasn't sure about. *I'm being paranoid.*

Latanya exited the Metro Subway station and confidently walked along the road. As she casually swept her hair back, she glanced behind her. Why was the junkie behind her? She picked up her pace.

Despite that, the deadbeat-with-a-noticeable-stoop kept up.

After 30 more yards, she stopped in her tracks.

Behind the 'junkie' was a beat-up black Continental. It fit right into the neighborhood. But the driver was the same man who had been sitting next to Jenkins, watching the fire.

They were planning to follow me! No way. Without looking at the car, or the short white 'junkie' following her a little distance away, Latanya promptly turned ninety degrees to her right at the intersection and walked straight towards the Walhealth store 20 yards ahead.

She entered the store and took a deep breath. Through the window she watched the short 'junkie' hustle to the car, which was now parked across the street some thirty yards past the Walhealth store. He opened the driver's door, and the tall man with the scar on his right cheek stepped out.

The tall man walked towards the store, and the short junkie walked somewhere behind the Walhealth.

Latanya strolled through the aisles. How could she shake them?

She looked at the door marked 'Employees Only." She glanced right, then left. No one around. She opened it and stepped in, almost colliding with the associate manager.

"Yo, miss, you don't look like no employee. Can'tcha read?"

"Uh, I, just got lost."

"Well, if you need to find somefin' special, the store clerk's gonna help you."

"Actually, I need your help, I'm being followed."

345

He stared at her for a full five seconds, sizing her up, judging. "Listen, we can't let just anyone go out this door cuz dey, like havin' a fight wit' their 'B' 'F' or somefin'."

"It's more serious than that. Could you just see if anyone is hanging out?"

The associate manager grimaced, looking her up and down. "Okay. Just don't touch anythin' here."

He went to the back door. He opened the door and looked outside. He quickly shut the door.

"Dere's a white dude with a black sweat jacket here."

Latanya grimaced and brought her middle right finger up for a quick chew.

"Look, I ain't got all day," the manager said. "Whatchu wanna do? Want me to call the cops. Or is dat the po-lice outside?"

She shook her head. "Never mind, I'll figure it out," she replied, leaving the office. Her head was throbbing.

As she walked through the cosmetics aisle, she looked out through the store windows. The tall guy was lolling about, hood on. She paced up and down aisle 'B', hoping for something. An inspiration, a knight on a white horse . . . anything. But nothing came, besides the realization that she would have to confront them. Police are useless, she thought, *It's all up to me.* She wandered down the cosmetics aisle. *But how?*

She strode into the adjacent aisle and glanced at the toys on the shelves. She spotted some plastic guns. As she reached the hair care section an idea took shape. *But first, no fingerprints.* She snapped up a bright red basket and went to the glove section.

Latanya found a box with 20 nitrile glove pairs in it. and ripped out the perforated cardboard top. She pulled out two gloves and put them on. She dropped the box into her basket.

Next, she went to the beverage section. Tonic, Cokes, Soda water. She found a still water. Raspberry-flavored. *It'll do.*

She returned to the toy aisle. After no more than ten seconds, she found it. It was an old-fashioned squirt gun, not the water-pushing hydro-cannons her young niece and nephew cherished when the temperature shot up to the nineties. This 'Squirt-master' was a modest piece, the kind she remembered from her

346

childhood. She ripped the two plastic rings that kept it attached to the bright cardboard backing.

She hadn't thought of doing this again, but they had left her without a choice. *I hate them. Why don't they just leave us alone?* First, she poured about half an ounce of the flavored water into it. She pressed the trigger once. Nothing. Twice, again nothing. But the third time, a thin spray of water drenched a Nerf ball to the right.

The associate manager passed the aisle, glancing in her direction. He stopped. She quickly placed the gun into the bottom of the basket, and pretended to look around at other things.

"Are you findin' whatchu need?" the associate manager asked, looking at her hands.

"No . . . I mean yes, I'm fine. Just picking up something for my nephew."

The manager nodded slowly, and pointed upwards. "We got cameras you know."

"I'm not stealin' anything." Latanya replied, her blood rising. She counted to three, then smiled at the employee. "Believe me."

"You best not." He crossed his arms, continuing his slow, irritating nod.

Latanya bit her lip. Just to make her point, she picked up another toy, some stupid action figure, and put it in the basket. She couldn't blame the manager, most likely woefully underpaid. This was a rough town. From what she'd read and heard, she understood that Harrisburg's hardest criminals were mere juvies compared to Bal'mer homeys.

The manager walked on.

Seeing that the traffic in this aisle of the store was low, she proceeded to place the basket on the third shelf. She once again pulled the little plastic stopper from the back of the greenish-yellow plastic pistol. She leaned the gun against the side of the basket, so that the water wouldn't escape. From her purse, she pulled out the self-sealing bag with the small vial. About an eighth of an ounce remained. She looked again to either side. Nobody. She also looked at the video lens thirty feet away. It would catch only her and the front of the basket. Anyone looking at the tape would think she was comparing dresses from the Beanie Babies' clones. Since she wasn't putting anything in her pockets, no one would care.

She carefully emptied the contents of the vial into the squirt gun and replaced the plastic stopper. Was it airtight? How fast would the cyanide eat through this plastic, she wondered, as she pulled the gloves off. She pondered discarding them, but didn't dare, should the gun have a leak of some sort.

She walked briskly to the check-out counter. She knew her testing of products in-store would appear odd, if not downright barbaric. Only the bottle of raspberry-flavored water and the action figure looked as if they hadn't undergone testing in the store.

Latanya handed the cardboard backing of the 'Squirt master' to the cashier. The barcode reader ting'ed.

"Had to try the thing. You know how that stuff from China is."

The store clerk shrugged her shoulders, as Latanya brought the last thing from her basket and put it on the counter.

With horror, Latanya watched as a small bead of water appeared on the left side of the stopper.

"That'll be thirty seven dollars and thirty-three cents."

Latanya nodded, perspiring, as she removed two twenties from her wallet.

She was about to say "keep the change," but that would have made her stick out even more from the five people waiting behind her in line. Already she couldn't seem to refrain from testing toys and rubber gloves in-store. She accepted the change. "Thank you," she said, suppressing the thought that someone might nonchalantly pick up and touch that drop of liquid in the basket. But then she sighed, calmly. In less than a minute, the lethal element would have evaporated.

The three items, along with the cardboard 'Squirt master' backing lay in the bag she stuffed into her handbag. She put on the right glove and picked up the gun in her right hand. She angled the barrel of the squirt gun towards the floor, so that no liquid would seep out the back.

As she walked out the front door she steeled herself for confrontation—if only she knew what form it would take.

The tall man was near the car. She sauntered slowly into the street until the tall man started approaching. She didn't run, and it wouldn't have helped her if she had; she couldn't have outrun him, even in the relatively low heeled boots she now had on.

The man stuffed his cellphone into his pocket, and approached quickly. He appeared huge, he was over six feet four tall.

Forrester held the gun behind her. To "load" it, she raised the barrel parallel to the street and squirted twice, behind her.

This could be unpleasant.

As he approached within ten feet, she held up the gun, aiming it at his scarred face. He registered surprise and he stopped.

The stream's trajectory went too far to the right. It whizzed by his left shoulder.

His expression hardened, as if he had just been called a "wimp."

Latanya adjusted. He was five feet away. She squirted again. Hit the left side of the hood, maybe the cheek. He didn't even protect himself. Barrel to the right. The stream hit him in the right eye.

"Goddamn bitch. Think yer funny?" he said, wiping away the liquid with his right hand. He put up his left in an attempt to ward off any more streams.

She jerked the barrel away from his hand and fired again. The stream hit him just below his nose.

He licked the top of his lip, then lunged towards her.

Oh no.

He brought down his fist hard.

"Thwack," followed by a sharp pain in her right wrist. The squirt gun skittered into the street.

The man grabbed her by her left hand. She lunged as far to the right as she could and lifted her boot over the plastic gun. But the thug jerked her at the last second and her boot stopped two inches above the asphalt, leaving the toy untouched. He started dragging her to the car. She didn't put up as much resistance as she normally would have. No kicking his groin, or trying very hard to get loose. Just some stopping, enough to communicate anxiety.

349

Using her teeth, she peeled off the glove from her right hand. With her right hand, she took the inverted glove from between her teeth and threw it in her purse.

The other hooded thug had already opened the rear door and was now opening the front one. He climbed into the driver's seat.

Her thug threw her into the car. Just before hitting the seat, she heard another car driving by and the sound of plastic cracking. Latanya felt relieved. No kids would be playing with the deadly toy.

"Move over," he said.

He was licking his upper lip again. *Good.*

The door shut. She touched her wrist, throbbing in pain.

The driving thug turned around. "Hello, Miss Forrester." His tone sent chills down her spine. A psychopath. "Where is he?"

"Why should I take you to him?" she said, biding her time.

The driving thug turned around, holding a big pistol with a silencer on it.

"We can just hang out until your brother shows up. And you—boom." He jerked the gun up. "What, 'dey don't see shot-up bitches around here?"

Man, they're good. Of course. Once again, Latanya was covered in cold sweat. She took a deep breath as she remembered Elaine's words. Three minutes if in the bloodstream. *I hit him in the eye. That must be fast, too.* "You win," she said, resignedly. "Turn left."

"Dat's a good girl." The driver put his gun down in the passenger's seat, and shifted the car into reverse.

"Turn right on this street. Left at the second intersection."

She turned to the thug next to her. He was rubbing his eye with his left hand. She leaned forward, peering into his jacket. She saw a pistol handle sticking out of a holster.

The car turned to the left. This street ran parallel to the one where the new safe house was located. But she could go as long as necessary, along these empty streets with rows of dilapidated buildings before taking a cross street.

"How far?"

"Go six blocks."

She glanced towards the thug next to her. He coughed.

They drove for another minute in complete silence.

350

The big thug coughed again.

She glanced at his face, then to the front, then back at him. After 10 seconds, he clasped his throat. His eyes widened.

"Turn left at the next street," Forrester calmly said, looking forward. Quietly, she reached towards the big thug's gun.

He grabbed her hand with his left hand. He was about to say something, but couldn't. He looked at her, his eyebrows raised. He stuck his tongue out. His saucer-like eyes searched to the right, then left.

With her left hand, she loosened his grip on her right hand and continued reaching for his gun. She grabbed the handle and gently pulled the weapon out.

He suddenly lunged forward, hitting the driver's headrest violently with his forehead. He jerked back. It was his last conscious action.

"John, what's wrong?!" The driver turned his head to his left.

Not receiving an answer, he kept turning around. Upon seeing his partner's face, he slammed on the brakes.

"I'm pointing his gun right at your heart is what's wrong, asshole. Keep your hands on the steering wheel," Latanya calmly concluded.

The thug swung his head back the other way towards Forrester. He acknowledged the gun with a slight nod.

"But I'll take you where you wanna go anyway. Just so you can see my boyfriend alive and well," Latanya said, coolly.

"Bitch. Waddya do?"

"Feminine mystique . . . not for you to understand. Now take the first left."

The driver glanced at his gun on the front seat.

"Don't even think about it. You begin reaching, I pull the trigger."

Ten seconds of silence followed.

"Now take a right on Etting," Latanya said.

The thug gripped the steering wheel with force. His eyes narrowed. But he had no choice. "Same street he used to be at."

How did he know?! "So you're familiar with the neighborhood. Good for you," she said, straining as hard as she could to appear cool, calm, collected.

The car proceeded, at twenty-five miles an hour, another hundred yards down the street.

"Now pull over at that boarded-up red house."

As he pulled over, his right hand fell from the steering wheel.

Latanya moved the pistol barrel three inches to the right, jammed it against the seat back, and fired. The seat muffled the shot, as the bullet hit the radio, tearing a one-inch hole above the second and third pre-set buttons.

The driver put his right hand back on the wheel, saying nothing.

"The next one hits your heart. And don't think I won't. As you see, I've killed before."

Bravado. But she could do nothing more. It was a stalemate now. She knew only that if he took his hands off the steering wheel, he would be dead. Maybe it wouldn't matter anyway. Maybe the thug would elicit no reaction. But maybe Paul would remember something else when he saw him.

If Michael had gone downtown, as he promised to do every other day, he'd have gotten the message. If he had forgotten or something, who knew where he'd be? If this thug got out, she could possibly shoot him, but even this part of Baltimore wasn't calloused to people being gunned down, and they could link her somehow to ballistics, couldn't they? Besides, she wanted no part in burying the corpse she had just produced. In short . . . she needed help.

"Honk the horn—"

The driver responded with a cough. Then another, this one violent. He brought up his right hand, signaling to wait a second. He started convulsing as he coughed.

Fumes of cyanide? How could that—"

Suddenly he jerked backwards, hitting the seat.

Out of the corner of her eye, she saw something move on the right, and glanced through the right front window.

The thug had already whipped his right arm over the back of the seat. He hit her wrist violently with his massive hand.

Pain shot up her arm. Her pistol fell to the floor. She grabbed her wrist, wincing. Latanya looked up into a set of neat grooves, winding through the inside of a very small pistol barrel.

352

"Now, you will do as I say. We'll get outta the car together," the thug said gruffly.

Suddenly, a loud tap on the right front window.

"You Dead!" she heard Michael say loudly and firmly.

She glanced and saw some smooth round attachment with a hole in it, on the end of Michael's pistol barrel.

The thug loosened his grip and the small revolver twisted down around his finger, barrel dropping down.

Latanya grabbed the gun off the man's finger. "Now unlock the doors."

Michael opened the car door. "That's a wise man. Stay cool."

Latanya almost let her body fall limp, from relief.

"Tan," Michael said, keeping his eye on the driver, as he picked up the other gun in the front seat. "What riff-raff you be bringin' here?"

"Mike, I gotta let Paul talk with the creep. He knows him."

"Sure enough. But let's take him inside first, before Grandma Jefferson across the street sees us with burners."

Michael turned back to the thug. "You retarded? You heard da' lady, now get outta da' car, and keep your hood on," Michael barked, as he pointed the gun at the driver's head. "And yo' hands up."

The thug got out, hood on and hands up, as told. He shut the door with his hip.

A cold chill went down Latanya's spine.

He was smirking.

She reached into the deceased man's right jacket pocket and took out the cellphone.

***

"Fuck!" Olson smashed his palm on the table. "No one could see it was Baltimore or D.C. she was headed to?! We can't cross state lines, but everyone else . . ."

By the time they had gotten permission to ask Maryland's Police to track her, the request to their State Police submitted and the latter responded by contacting Baltimore . . . she had long been off the train.

353

Olson picked up his phone and dialled a number in Philadelphia. After one ring, he hung up. *No Feds, not yet.*

53

As soon as the thug's hood came off, a flood of memories overwhelmed Paul. In an instant, visions appeared of the first attack from the black sedan. Then the last conscious seconds of Hollidayton. He felt blood leaving his cheeks.

"Honey, you okay?" Latanya asked.

"Yeah, yeah," he responded, taking in a deep breath. "In fact, it's great you brought him in. Let's tie him up and figure out what we do next." Paul circled him, examining for details.

The man smiled. "Surprised to see ya' ain't the only one wit' the luck of the Irish?"

"Shut-up. This isn't over yet," Paul said as he kicked the man in the back of the knees, forcing him to kneel.

After finding and removing two pistols, a Taser stun gun and a knife, Michael and Paul secured him, using some electric cords from lamps left by the previous inhabitants.

Paul approached Latanya and lowered his mouth to her left ear "This guy is key. He might lead us to the mind behind this."

Latanya turned to her brother. "Mike, we're goin' upstairs to discuss this. Could you keep an eye on him?"

"Sure, but I ain't waitin' for you to do no whoopee. You best be doin' some first-rate plannin', 'cause I need to get ma' life back, you--"

Michael!" Latanya gestured with a 'time-out' signal. "This guy is a big fish. He knows something. We'll be back down in a few minutes."

Michael cast his glance downwards. He pursed his lips and nodded.

Once upstairs, Latanya took Paul's hand in hers. She put it to her chest.

It was the wanton frightened look in her eyes. He just had to kiss her. He brought his lips close to hers.

She reciprocated, bringing her lips to within a half-inch of his. She stopped, "No, not now. God I want to, but not now, sweetie."

"I, I understand," he said, shaken.

"The police know. They were following me, I'm sure."

"So they are in the know, huh? Finally," Paul said, answering his own question, not waiting for Latanya's response.

"Yeah. No turning back now. So Paul, what did he do that night, and what do we do next?"

Paul gave a synopsis, but before he could finish describing the last minutes in Hollidayton, he heard a pot hitting the floor downstairs. This was followed by the sound of glass breaking.

Paul had reached halfway down the stairs when he saw Michael, his hands struggling to free two thighs wrapped around his neck. Michael's eyes bulged grossly.

Paul lifted his HK and fired at the thug's calf farthest from Michael's head.

Blood hit the cabinet door beneath the sink.

It had no visible effect on the thug.

"Hey, Italian! Next shot to your neck, and I don't miss nowadays."

The thug's legs loosened, and Michael's head slipped out from under the thighs. Michael coughed twice.

The thug kicked Michael in the head.

Michael scooted away, took five deep breaths and shook his head. "Man, that guy is psycho. Can *anyone* tell me why we're keepin' him alive?"

"Yeah, Michael, but I'm quickly forgetting. For now, Mr. Badass has some info he needs to share with us. In the meantime, Tan, could you get me the gauze, please? If we set him back on the streets, he can't be completely bloody, just a bit.

To Paul, the Italian looked complacent, bored even.

"Listen" Paul said before pausing. "What's your name?"

"Snoop Dog."

Michael smiled, rubbing his neck. He took two big steps and laid into the man's stomach as hard as could with a solid right kick.

356

The thug, on the ground already, doubled over.

"Good, Snoop," Paul replied. He turned to Latanya. "We'll need more gloves."

"Sure," she replied, approaching the counter and jerking a pair out of the box she had put there a few minutes earlier.

Paul put them on, tightening them with a satisfying snap. "Get up!" He yelled at the thug.

The Italian stayed down. Paul handed the gun to Latanya and then walked over to him. He stopped just behind the killer's head. He threw a solid punch in the man's face, connecting with his left cheek and nose.

The man's head jerked stiffly.

Paul then grabbed him by his ears and started lifting.

The Italian said nothing but scampered to get up. *As painful as it looks.*

Paul sat the man down. When he was sitting comfortably, Paul cupped his hand in back of his neck and threw him out of the chair.

His forehead hit the floor and he tumbled on his side.

Paul was pleased with the effortlessness of the move—he had developed some nice pecs over the past two months of seclusion. He reached into the killer's back right pocket and removed a wallet. From the wallet, he took out the driver's license.

"Francis Antoine," he announced. "Listen, Frankie boy, here's what we're gonna do. We're gonna ask some questions. If you answer them truthfully, we spare your life. If not, then we're gonna shoot you 'til you die. And we're not doctors are we, Mike?"

"No sirree!" Michael replied, rubbing his temples.

"So we don't know where the kidneys or liver are. You might die quick and you might die slowly. Like how long does it take to die if your balls are shot off, do you know? We don't, do we Mike?"

"No idea."

The short stocky man remained expressionless. Only his eyebrows moved, slightly upwards.

"So, we'll start easy. Who's behind the voting machine manipulation? Is it Danabold?"

357

"Don't know what you're talkin' ab--"

"The company."

The Italian shrugged.

"Is it Leboeuf?" Paul asked.

Latanya looked closely at the ruffled man's expression.

"Marty Leboeuf, living in Cumberland." Paul continued.

"Don't know no Leboeuf."

"You don't know Marty Leboeuf? Okay, who do you work for?"

"Motha Teresa."

"Cute. Michael, could you soften up this hard guy, fast?"

"Sho' thing," Michael replied, as he got up. With one lightning fast movement, he hit the thug on the nose with the butt of the gun, producing a loud 'crack.'

"Ow, assholes. You Fucks." Antoine murmured, blood pouring all over the front of his track suit. He squinted, with bloodshot eyes, at Paul.

"Right. So now that we've got each other's I-D's straight," Paul continued, "you better spill all, because . . . the nose is really, really sensitive, and you're gonna leave here looking like the Mummy."

"Eat shit," Antoine responded.

Paul forced himself to calm down. "Okay, tough guy, before you get all scatological on me, once again, slowly. Who do you work for?"

The thug squinted his eyes further, his irises a deep black now. "Da Catholic Church."

"Doubtful. Even that institution has a bit of a conscience. Michael, let's tweak his shnozz a little."

Michael nodded, took hold of Antoine's nose between his fingers and started moving it side-to-side.

A tear descended down the thugs nose, but his face was granite. He struggled to get out of the cords. They just cut deeper into his wrists.

"Michael, I don't think he likes you being here," Paul said. "So, I have an idea. Go take his and his partner's phones and walk them around the neighborhood. Then stop in front of one of the meth labs you'd told me about. You know, in case his friends

358

from the church are tracking it. Oh and then call the three numbers he's been calling most. Just say 'hi, this is Francis and I'm in trouble. You know, with that cute accent of his."

Michael smiled. "Sure dude. Then I'll take out the batteries."

"Yep. Oh and forward those numbers to me first, for safekeeping."

"See ya later."

The door closed. Silence followed.

"Jenkins brought me in on 'dis," Antoine finally said, in a defeated voice.

"Who's Jenkins?" Paul shot back, to make sure they were talking of the same.

"Some ex-military contractor."

"Hmm. Okay, that explains who brought you in . . ." Paul tapped his chin, "and do your work for him now?"

"I'm doin' it in his honor."

"Is he dead? How noble," Paul said, turning to Latanya. She shook her head.

"But that's bullshit," Paul continued. "Tan, he wants us to play with his nose some more."

"Yeah, yeah, it's this guy from the computer company," Antoine answered, quickly.

Paul stroked his chin. "Then is it Marty?"

"I don't know no Marty." Antoine's eyebrows lifted up slightly, but he kept looking straight.

Paul glanced at Latanya. She shook her head. Paul couldn't control himself any longer and brought the edge of his left hand down on Antoine's nose.

Antoine emitted no sounds. He looked down, then up, smiling.

Paul broke out in a sweat. He turned away. "What about Lewis?"

Antoine shrugged his shoulders. "Who's that?"

"He was the one who lived next to Kendrey, in the apartment building."

"I don't know any Lewis's."

Paul studied his face intently. He glanced at Latanya.

359

Her forehead was furrowed.

"So you know Lewis. The tanned man. We'll come back o him. . . Okay. So what happened that day? You ended up walking away, I see."

Antoine told them the story up through the moment the police officer appeared in the bedroom doorway.

The story appeared logical, except that Antoine came out innocent. *I'll leave that for now, too.* "And what happened next?" Paul prodded.

"You were shot."

"Who shot me?"

The Italian hesitated. He was contemplating something.

"I dunno. You knocked me out, remember?"

Paul turned to Latanya and nodded. He reached over to Antoine's nose.

"But it could've been Parkett, the dude that came through the door," Antoine let out quietly.

"Did Kendrey fire back?"

Antoine stared back at him like he was nuts. "Da computer freak? Naw, he just stared."

"What happened to him?"

Antoine looked at Paul for a full five seconds. "We took him, they took him outta there."

"Where is Kendrey now?"

"I dunno."

"Where'd you take him?"

"Where we took everyone, all the bodies too, to some rundown building on 3ʳᵈ Street in Harrisburg."

"What happened there?""Nothin'. They took the corpses n' put 'em on plastic sheets. Gave me some first aid. I dunno what happened to da computa' guy. . . looked dead to me. He wasn't important, once we had your phone."

Paul stared at Antoine for a second. *You are so lying to me . . . but I can't tell on what.* "Who gave you orders, then?"

"Jenkins. He broke his hip that day, but he was still givin' orders."

"Okay, that's clear. What isn't clear is where Parkett is today," Paul resumed.

"No, dat's clear, too," Antoine said, with resignation in his voice. "He's in hell."

"Don't be fun . . . "

"Heaven? I don't think so . . . he's in the backseat of the car."

Paul rubbed his chin. "What happened to Jenkins?"

"He O-D'd on the blue-greenies last month. Too much chasin' pussy."

Paul couldn't help smiling, and glanced at Latanya, but realized he had to change his demeanor instantly. Poker face.

"Let's not forget we're in the company of a lady," Paul said, curtly. He circled the Italian once. "Now tell me about Lewis."

Antoine fidgeted. He glanced up at Paul and then down at the ground. He shrugged his shoulders.

Paul sensed it was more. "So it's Lewis that set this all up."

Antoine looked forward, saying nothing.

"Lewis, Lewis, the man with a spectacular tan. I should've guessed," Paul ruminated out loud. "Of course. He was living in the next apartment. Responsible for the project. Backing of the company, too. So the resources. Yep, I suspected, the helicopter."

"Anybody can hire a fuckin' chopper these days," Antoine interjected. "Listen, kid, you're outta your league. This is big, and I'm not cheap. Let me go, and I'll ask them to spare ya' life. And my guys are lookin' for me in the 'hood. They'll find me soon."

Paul scratched his head, ignoring the statement. Needle in a haystack.

"Ya, Antoine, a good deal, but nyaa', I'll pass. Here's what we're gonna do . . ." Paul started, using a school principal's tone of voice, "you go back and tell your boss, whoever it is, I wanna meet with him in three days' time, on Friday morning."

Antoine shrugged his shoulders.

"But that's not all. Listen carefully. I want him to bring me proof of Kendrey's death, and three hundred-fifty thousand bucks."

Antoine stared at him. "And what's in it for the boss?"

"I stop investigating. And I don't tell anyone about the helicopter at the Stoltzfuss farm."

Antoine lifted his head up. "Why?"

361

"I'm a human interest reporter. I report on farming equipment, warm fuzzy bunnies and Olympic medal hopefuls." Paul looked up and sighed, "I want my life back and some money as compensation for this crap I've gone through."

"Dat's a lot of money for a chimp-reporter, for . . ."

Paul held up his hand.

"You know it's not. Danabold, or whoever, will get to continue making defective machines, saving them tens of millions of dollars. The programmers will continue to program incompetently, et cetera, et cetera. In fact, double that amount. I want seven-hundred thousand, so I can take my sweet time finding another job." Paul looked at his fingernails. "Now, about the meeting. It'll be at 7:00 a.m. somewhere around Harrisburg. I haven't decided where yet exactly," he said, stroking his chin, "but -I'll S-M-S you Thursday afternoon . . . on Parkett's phone."

Antoine nodded ever so slightly.

"Oh, and Antoine, I just wanna warn you, if you guys bullshit me, the deal's off, and for your information, I studied forgery. If you kill me, then ten people are going to send off my account of what happened that day to the police. It names names, and now will include yours, along with phone numbers."

Antoine nodded again, but Paul registered the beginnings of a grin. "Sure."

Paul tugged Antoine's red-splotched sweat jacket upwards. Antoine got up.

"And forget about finding us. We're gonna move from hood to hood, city to city. Let's face it, if the Pennsylvania State Troopers and Baltimore police couldn't find us for three months, then neither can your punk operation."

Paul thought he saw Antoine break into a wide smile.

As the three of them walked down the hallway, Paul lifted his hand. "Whoa, hold on a second, everyone. Let me check his partner out. Don't want this man coming back in five minutes trying to kill us." He handed his gun to Latanya.

***

362

Michael joined Paul outside. "Good idea, Paul, was thinkin' the same." Michael stepped around the car and opened the left back door.

Paul watched Michael through the window as he removed a pair of brass knuckles, a butterfly knife and a small gun from an ankle holster.

"This dude liked his toys. Where are the transformer dolls?"

"You might wanna check the trunk, too," Paul said as he opened the front passenger's door. He dropped the glove compartment door and took out a .357 magnum.

Michael opened the trunk. He removed the top of something that sounded to Paul to be wooden. "Got everything fo' a par-tee."

Paul walked to the trunk and saw, in a polished wooden box, two grenades, some red and white wires and something that resembled the clay he used to ply into little lions when he was seven.

Michael took out the case and Paul shut the trunk. They entered the house carrying the wooden box. Paul glanced at Antoine as they walked past. He shrugged.

Michael stopped. "Wait, dere's one mo' place."

This time, Latanya handed Paul the gun, and followed her brother out. He opened the driver's door and flipped down the sun visor. There was nothing. He then reached down, deep under the steering column. "Mm-hmm," Michael mumbled. Michael pulled off the tape that had held the .22 caliber in place and showed it to Latanya.

A cold chill descended Latanya's spine.

They re-entered the house.

Michael pulled out the .357 Magnum. "Paul, this guy's just full of surprises. How about we strip him down, in case he starts wantin' to play James Bond."

"Good idea . . . Latanya, would you cover us?" he asked, gesturing for her to take the gun.

She grabbed it.

Paul strolled to his own coat and from the left pocket removed a new Leatherman Wave Michael had procured for him a few weeks earlier.

Paul worked on the upper body clothes, while Michael cut through the Italian designer slacks with the scissors and the razor-sharp knife attachment.

When he finished, Paul stood back. "Nice little paunch you got there, Antoine. Some women find that cute."

Michael chuckled. "Yeah, definitely in this 'hood."

Antoine said nothing, but his eyes seethed at Paul, his eyeballs bulging like a Maori warrior's. "You fucks," he said calmly. "You had ya' chance. When the time comes, I'm gonna make it hurt."

Paul took a step towards Latanya and signaled for the gun. She gave it to him. He turned nonchalantly and fired the pistol. Again, no sound besides a muffled 'Phutt.'

The left wall of the corridor was splattered with crimson-colored drops. Where the Italian's right ear had once been, fleshy red cartilage dangled.

"Aargh!!" Antoine yelled, gritting his teeth. He dropped to the ground on his knees.

"Now listen good wit' ya' otha' ear," Paul said, calmly, imitating Antoine's accent, "Cuz' I am one crazy mad muthafucka." He spun the gun around wildly. "You come back here before Friday, and the deal's off. Oh and then we drop you and your partner's phones at the police and . . . " he paused, grinning "these don't look like disposables."

Paul nodded to Latanya.

She left and quickly returned with tape and gauze. She bandaged the thug's ear.

"Go ahead, Michael, set the man free."

Holding a small pair of wire-cutters, Michael stepped in back of Antoine.

After a few clicking sounds, the thug brought his right hand up to his right ear, glaring at Paul.

Latanya handed the car keys to Antoine.

Michael pushed the door open.

The three watched as Antoine, dressed only in underpants and socks, with bandaged left ear, walked briskly to the driver's side of the car, got in and drove off.

As soon as he was out of sight, Michael turned to the other two. "Okay, folks, time to move our butts."

"Right-o, Michael," Paul responded. "I hope it isn't far. That crockpot weighs a ton."

"Whatchu want, Paul, a mule? No, we 'da modern-day nomads. Besides, without this, I wouldn't have any more of your crappy chili!"

Everyone joined in his laughter.

Paul sighed deeply, taking in the fresh air that had wafted into the flat. It felt like a 500-pound weight had been removed from his shoulders.

Michael stretched his arms out and shook his head. "I'll be back in three-quarters of-a-hour. Gotta fix dat wiring and throw some more shit out the back," Michael said, anxiously. "Gimme the crockpot." He took the pot, threw in some tools and headed to the front door. He turned around. "Sis', dat's forty-five minutes."

Latanya shook her head, smirking. "I gotcha, Bro', I gotcha." She smiled widely, skipped to him and brought her arms around his neck, forcing his head to sag. She planted a big kiss on his cheek. "Thanks, Mikey, for saving my life . . . and for all you're doing."

"We're Forrester's, Sis. You just get that 'P' H' 'D', you hear?" he said, as Paul opened the door for him.

Paul and Latanya turned towards one another. Paul swam in her brown eyes for a full ten seconds.

She moved her lips close to his. He moved his face down and revelled for a fraction of an eternity in her soft sweet lips, sensing it would be the last time ever.

She jerked her head back. "Oh God. What torture."

"I know. We can't, not now," he whispered, resigned.

"Not unless . . . we want to die like this . . ." She cast her gaze down. When she looked back up, seconds later, her eyes were misty. She squeezed his hand, hard.

Latanya departed, once again ignorant of their new address.

54

"They think Marty's in India," Elaine said.

"In India?!" Latanya responded.

"From the neighbours' description of it, not voluntarily. The missus shed a lot of tears. The people living there now said they left a month ago. So I called the company. Told them the same I'd told the neighbors - it was a follow-up on a medical lab payment. The 'H' 'R' department said he'd been transferred."

"Weird. But who knows, Chennai is a future high-tech hub, right?"

Elaine shrugged her shoulders. "Not the world we grew up in anymore. Anyway, here's the Danabold Annual Report you asked for."

"Thanks," Latanya said, stuffing it in her bag. "You know, I wonder just how dangerous it is for Paul to be doing this. I mean, according to Michael, he just paces back and forth, and then writes another circle on the wall, with a name inside. Says he's been looking like the nutty professor with a gun fetish for the past two days."

"It'd be worse for him if he didn't."

***

"That just doesn't make sense," Paul told Michael, "unless he orchestrated the whole thing . . . and now sees it crumbling around him."

Paul paced back and forth. Lewis came to mind, but then he stopped in his tracks. His line of reasoning was wrong. Was Danabold behind the whole thing? It would be easy to get caught. Such a high profile project. Would a company risk its entire business model on a political campaign? For what? If a plot was uncovered, they'd lose credibility, get sued for sure. At the very least they would never have this revenue stream again.

He had found, sure enough, that the voting machines were a small component of their corporate empire and growing only marginally. *Why create a potential company-destroying scandal, for a tiny part of the revenue stream.* Something wrong with the scenario.

Latanya, at his request, also found out that Lewis was still with the company. *Access to the apartment, knew where I was, and probably put a tracking device on Kendrey.* And who would have done this but a small group of fanatics—right wingers—with connections. *From where else the badges in Hollidayton?* And the police had stopped their investigation cold. Then again, what wasn't possible to do with modern computers, with some morally rudderless geeks at the keyboard.

Jenkins. Latanya said he had been called into small rooms to train others on rallying college kids. Paul shuddered to think of the consequences.

***

"So, what have you got?" Olson looked like he had just stepped out of a rotating dryer.

"Paul is alive and well."

He smiled, nodding. "Figured as much. Where is he, D.C. or Baltimore?"

Latanya nodded in acknowledgment of Olson's tenacity. But she still wanted to hedge her bets, so she chose to tell a white lie. "I don't know how he got there, but he's in Baltimore."

Olson raised his eyebrows. "The address?"

"Don't know."

Olson's face grew red. Veins on his temples seemed to expand. He crushed the empty styrofoam cup in his hand.

"Lower West Side, swear to God that's all I know."

His jaw stiffened.

"But listen. This Friday, he's going to be in Mechanicsburg," Latanya quickly interjected, intimidated by the display of anger from the normally placid man.

"Why?!"

"He'll be meeting with the organizer of the whole shoot-out in Hollidayton."

Olson nodded. "Look, Forrester, this sounds nice, but I don't believe it—especially that part about you not knowing where he is now. Let's face it, your boyfriend's still a fugitive. I could have the Baltimore police swoop down and find him by Thursday." Olson re-crushed the cup, again generating the unpleasant squeaking sound. "And you, you - you're not getting any more chances from me,".

"Probably you could, detective Olson. I can even give you the house address where they were yesterday. They said they might be moving uptown," she fibbed, "but you won't get the guy who set it all up. And I'll tell you a few things now. I saw Paul this week. I talked with him, it's true, and I even gave him your number, so he could turn himself in. But it's also true that I don't know where he is because he did that on purpose. In case *you* interrogated me . . . or," she paused, "the bad guys did. So you can water-board me, pull out my nails and run my feet over hot coals, I could *not* tell you where he is now."

He leaned back and looked at her.

She knew he could come to only one conclusion.

"Okay. What's the meeting gonna look like."

Latanya related the location, the time, and Paul's plan.

"I'll think it through. Let's meet tomorrow morning at seven. He leaned back and scratched his head. I know that place like the back of my hand, but we're still gonna need a day to put everything in place. I wanna call Paul."

"Fine."

"Oh, and if this is a set-up, Forrester, you're going to jail, and not in Hollidayton, either. Where I'll put you, there won't be anything 'holiday' about it. Just to make sure you understand we mean business this time . . . we're gonna lock you up on Thursday afternoon, on suspicion of harboring a fugitive."

55

It was Friday morning, and the only life Antoine could discern, from his vantage point in the middle of the Mechanicsburg amusement park – where booths once sold popcorn, cotton candy and the opportunity to throw oddly-weighted bean bags through holes in clapboard walls – were eight foul-smelling forms, huddled under brownish-gray blankets seventy yards away. It was 6:03, almost an hour before the meeting, and he had done a cursory inspection this morning already, not twenty hours since they had so carefully patrolled the place yesterday, finalizing positions for his crew. He had also scanned the place for electronics. There was nothing.

I want his hide like a fat man on a diet wants a Philly cheese steak sandwich. His face looked a wreck in the mirror this morning and was still throbbing with pain. Sure, he had swung by the decrepit townhouse a couple of hours after he'd been so brutally humiliated. It had taken him forty-five minutes to hook up with Girelli and Jameson and another fifteen minutes to rustle up a set of clothes that fit. When they finally got there, the apartment was empty. Damn, why did they take away phone booths . . . could've watched the house from one of 'em and guided Girelli to the place, he lamented.

And now, he would've liked to have Parkett nearby. He hadn't offered an alternative explanation to the autopsy results of heart attack at the hospital. If word got out of something else, the story of his incompetence would abound in his corner of the underworld. As it was, there were no witnesses. No rumors about a single, petite black woman who might have rolled him and his partner. The hearsay alone would not only have ended his career, but put his life in danger as well. *Soft* is what they'd call him—or "past his 'use-by'" date. But a natural event, that was a different story.

So Antoine stuck to that story. Brave Parkett had suffered a fatal heart attack—and to make matters worse, it happened just when Antoine needed him to cover his flank. A local internist, an acting coroner near York who was short of funds for a small yacht, confirmed it.

But that was all behind him now. He had two of the best weapons men in the business here, one of whom was Girelli. The guy could hit a quarter at fifty yards without aiming. And he had just found two good vantage points for them.

As to these homeless people, they smelled. Jameson had repeated three times that they'd been kicked off the night train the night before, some sixty yards to the right, but he had to make sure now.

Antoine approached them. They were snoring and when he shone a flashlight beam on one of them, the chiseled gray-bearded black face just looked up with a note of fear and pain. The bloodshot eyes testified to so much drinking that the man's liver was probably gone. He rustled up the second one. White, about sixty, missing most teeth. The third one was younger, black but graying. He burped. Antoine roused the fourth one and that one, also black, jumped to his feet. He tried to push Antoine.

Antoine side-stepped and hit the guy in the back of the neck with the butt of his gun. "Why do you bums always have to pee on yourselves? Smells like a goddamn outhouse here . . . a sadder excuse fa' humanity I never seen." He pressed his forearm sleeve under his tightly-bandaged nose.

Since his men would be using silencers, this pile of drunks wouldn't even hear what was going on. And should one get in the way, he figured he would just dump the guy's bullet-riddled body in the fast-moving river forty yards away, along with the other one he was supposed to produce. They were notoriously undependable witnesses, though, so he had nothing to fear from them actually seeing what transpired.

As he walked back to his spot, Antoine wondered what the beaches on the Eastern coast of Italy looked like. He was partial to rye, but he could get used to grappa. Would they have good ones in Azzaro? With three million dollars, he would buy a house on the beach and a stock it with a bevy of young babes. Which beach? He stopped his mind from wandering along that beach. *I'll celebrate in two-and-a-half hours, not before.*

370

Then again, things were looking good. Chimp-reporter was still a fugitive, so the guy wouldn't be wanting the Law around any more than he did.

The only outstanding question was how much time to give for the information exchange. He decided he'd just wait for the signal. And if anything got out of hand, he could pull the plug on everyone. Except Girelli, of course; the guy's family in Italy might pay him an unpleasant visit, Antoine reasoned, if he were killed.

That Robinson was terribly naïve. Antoine'd heard the argument before about telling everyone "if something happens to me." The conundrum was that either he had left the files with people who've studied them already and knew, or he was bluffing. Antoine was sure he was bluffing. The only one who knew the story was Latanya Forrester and her brother, and he knew where to find her. Eventually, he would find Michael as well, *whenever and wherever he comes up for air*, it would be the last piece of business he'd be taking care of before flying off to Italy.

There were two entrances, rear and front, about three hundred yards apart. The reporter would be using the front entrance, no doubt. There would be a possibility of witnesses. Then again, if Robinson had chosen this place, it was because he wanted confidentiality. Unlike on the weekend, when the flea market was in full swing, this place would be crypt-empty on Friday.

Jameson he put in the highest building, which covered two-thirds of the park. Bueller would be mobile – a sweeper. Girelli would be in the upper tier windows in the octagonal wooden structure, halfway between the kiosks and the rollercoaster, which once sheltered the bumper cars.

He himself would be covering the Eastern side, with an unobstructed view of the railway tracks eighty yards away, from a nondescript little trailer, used to grill and sell Italian sausages. He could smell the aftermath of fried onions and grilled meat from the previous Sunday, as he stepped next to the trailer stand. Once he lowered the panels, he'd also have a clear bead on the whole area near the gazebo where Robinson said he'd be.

But to make sure he'd do that only at the last minute, he installed wide-angle peepholes in both the front and back panels of the trailer.

371

On the other side, the "front" of the stand-on-wheels, was the building with two shops. One promised to "Buy your gold for cash." From what he could see through the scratched and dusty windows, the other offered souvenirs from trivial events and junk masquerading as "antiques."

All in all, it was good. Even if Robinson lied and showed up in the stores, or even in the long booth behind him, his men would just have to turn a little, their line-of-sight still unobstructed.

***

Paul could barely breathe.

"Man, you stink worse than ma old dog after a swim in the sea."

"Just a lit-tle longer, Mike, and I will be a distant memory."

"Survived two attempted killings and now I'm gonna die from asphyxiation. Next to a baldin' white guy. Jes' ain't fair."

"I know, but Rogaine doesn't work on my hair-loss pattern."

"Ha, ha, so funny, gonna bring it up at my 'Welcome to Hell' Party."

The impromptu hiding place was no more than a box big enough for two, with twenty-five straws providing oxygen to them. Paul had known about this storage cave for the booths for ages. He had played in one for hours when he was five, when Pa had brought him here as a reward for accompanying him to the doctors and health insurance administrators in Harrisburg.

But now that they'd been lying in it for six hours, Michael was right . . . it smelled like the bad end of a Pottstown pig sty.

Paul focused on the screen he had just turned on.

"One man is in the Octagonal building, Michael. And our friend Antoine is due South, but I'm not sure where. He disappeared behind the sausage cart."

"Okay."

"Ready?"

Michael nodded.

"It's showtime."

***

372

By 7:00, the mist lay heavy in the air. Antoine looked through the peephole in back of the trailer. The sun was just coming up. Antoine heard a car engine in the distance. It stopped about a quarter mile away.

Through the peephole he watched his client walk slowly across the grass towards the three booths. He was alone, carrying only the oversized green satchel. As the gaunt man approached within thirty feet of the store in back, he stopped, slightly left of the stores, looking out onto the small gazebo on the other side of the narrow, fast-moving river.

"Robinson?!" his client called out.

Antoine heard no response but some birds twittering away.

The man turned towards the booth on the left. "Paul Robinson?!" he repeated.

Antoine heard an explosion and a loud creaking sound far behind him. It came from a point outside the view of his peephole. He looked through the crack on the panel.

A gray wooden roller coaster car began its descent down the track. As the car reached midway down the first "hill," it left Antoine's view.

Antoine thought to unhinge the clasps holding the panel in place, and throw the panel out, but stopped himself. *Girelli and Bueller'll have the bead on him.*

The roller coaster car crashed into something with a loud crunch, sending, from what Antoine could see through the crack, dust and chips of wood in all directions.

His client ran a few steps in the direction of the roller coaster.

Between the crushed and fallen timbers, about seventy yards to the left and behind his trailer Antoine saw him – Chimp Reporter.

"Yes?" the reporter responded.

The client turned to the left to face him.

The reporter wore what looked like a pair of beat up Ray Ban tortoise shell sunglasses from the 70's. That classic eyewear, along with Robinson's forehead and nose bridge, were all Antoine could clearly see from his vantage point.

The reporter appeared quite still. "Sir, this is quite a surprise," he finally said.

373

Antoine raised his Heckler and Koch, behind the removable panel, and took off the safety. *Not an easy shot, but not impossible, either.*

The client shrugged his shoulders.

"You son-of-a-bitch," the reporter blurted out.

"A bit harsh there aren't you?"

"Don't think so. Five good, innocent people dead. Hundreds of lives disrupted, and a perversion of this country's sacred construct of voting."

His client waved his hand, as if swatting away a fly. "Details, details. Millions could die, if Iran gets nuclear weapons. So this five, six actually . . . doesn't it pale in comparison? Isn't that a small price to pay for the preservation of American values?" he said loudly, glancing suddenly to his right. "For being able to continue building our military, without interference from liberals, for allowing corporations to do what's right for the country once and for all. Remember who John Galt is?" He paused, waiting for a response.

"Someone with part of the answer . . . for high schoolers."

"The whole one. But let's not split hairs, Robinson. I understand you wanted to know something more specific."

"Yep . . . Joshua Kendrey, or maybe I should call you, Joshua Benedict Arnold. Was there a glitch in the computer, or was it a set-up?"

"It was neither, Paul. Why do you want to search for demons? The program was just tough for some old people to get." He shook his head and, from what Antoine could see, stiffened his jaw.

"Bullshit, and you know it . . . Did you shoot me?" the reporter asked, almost expressionless.

The client stood still. Then he laughed, loudly, full-heartedly. "Paul, Paul. Don't think of it so badly. After all, you survived, didn't you? . . . Most participants didn't."

"So, how much did you get from the company?"

"Alas, ye of small minds-ville. Just my bonus. Why would I get anything else?" he asked, his voice on edge.

"To keep quiet."

"About what? As if I need money."

"Yeah, I researched that too. Your family is what, Pittsburgh's 9th richest? And has huge holdings in Danabold. But tell me something else. You say the machines are in order."

Kendrey glanced back towards the sausage trailer. He was perspiring. "Yes, absolutely."

"Then why the three districts, voting abnormally Republican . . . Senator Dooley received eighty-three to eighty-nine percent of the vote . . ."

"People coming to their right minds." He laughed, but this time, the laugh had a nervous edge to it. Then Antoine noticed it. His client's hand was shaking.

Robinson shifted his weight. "You, as it turns out, are normal. So tell me. Why? Did your father abuse you as a child or something?"

"Paul, 'Chimp-reporter,' as my new colleagues call you. I think this conversation has hit a new low. It's over. Should have stuck to your strengths. Not political reporting."

Antoine expected a denial from Robinson, but he held his head absolutely still. Maybe a stiff neck? Strange. *Where did he walk in this morning?*

"Fine, I guess that takes care of the details," Robinson muttered. Now, what about my seven hundred grand?"

Finally. Dat idiot's gettin' to the point. Antoine lifted his gun to just below the counter, from where tomorrow some slob would be handing stale sausages to the sorriest people in the county.

Kendrey held out the thick computer bag. "The ones that would be in here?"

It took Antoine less than three seconds to unhinge the panel, throw it out, lift his automatic pistol with sound suppressor and aim it squarely in the middle of Robinson's face, between the wood and rusty wheels seventy yards away.

Robinson didn't see him; he kept looking straight ahead at Kendrey.

Antoine pulled the trigger. He thought the bullet hit Robinson in his upper left forehead, a bit high. He just didn't have a clear bead either, so he wasn't sure.

Another shot rang out. It hit Robinson in the temple, judging by the unnatural jerking of the head. Blood spurted out in a hefty

375

stream. *Good shot Girelli!* The other tore into his neck, sending another spray of blood out. The right side of his face sagged, and he shifted to his right, but he remained standing. *Odd.* But maybe he had been leaning against something.

Kendrey pulled out a Smith & Wesson and fired a shot. The recoil almost sent the pistol out of his hand.

Miles off. Turkey's never seen a gun in his life.

Antoine set his Heckler & Koch on full automatic, aimed between two wheels and then emptied nine rounds in three bursts. Antoine saw the wheels rip but something behind them too, blood spewing everywhere. The reporter's nose was destroyed by other shooters. *But he's still standing?*

There was nothing left of the face . . . except a black line.

He glanced towards the bums. *A wire?*

***

Paul pushed the button on his control pad, shifting his double's weight back to his left. He had fashioned several dozen mechanical scarecrows during his teenage years and in his early twenties. He found it pleasing to outsmart the voracious birds. But never had his life depended on the simple mechanical device, made of flywheels, a short crank shaft and a simple hydraulic control, to show him realistically shifting his weight. He also positioned twenty separate plastic bags with a deep scarlet-colored corn syrup mixed with flour, under the mask of the 'double.' This, he had to admit, was the most complex version of the scarecrows he'd ever made.

The mask was the toughest thing. It had taken Latanya a while to convince the forensics prof at Penn State to show her how to make one.

Paul noticed a nervous edge to Kendrey's voice. Although he lay, smelling like urine and booze, fifty yards east of the contraption, Paul could tell through the microphone and on the screen that Kendrey wasn't himself. He lowered his mouth to the small mic, connected by wire to his 'double' in the booth. "Ouch, that hurts," he uttered.

***

376

Suddenly, before he could get a bead on the bums, Antoine heard a different gunfire bursting, staccato from behind him. Antoine turned around and kicked the panel out of that side, ripping out part of the pressboard frame.

Jameson was no longer in the tower two hundred yards away.

Antoine saw a man running at the tower's base, wearing what appeared to be a cashmere coat. *Coverts!* He aimed at the stranger. Pulled the trigger. Late.

The target disappeared behind another building. The ricocheting rounds sent chunks of old stucco flying. *Where the hell did that guy come from?* They had carefully scanned the grounds for any electronic signals. *The SOBs must've come without their phones and hid out on the grounds two days ago!*

Antoine turned back around. He redirected his rage at another plainclothes cop he saw seventy yards away, approaching the abandoned bumper-car building. *Girelli is up there.* Antoine aimed and fired at the man's chest.

The man went down.

Cashmere Coat'll be here soon. Antoine quickly swung back around to the other window. The man appeared for a second from behind the gold-buyers' shack.

Antoine aimed. A two-burst shot. Man hit in the arm.

Magazine empty. He jerked it out, threw it on the floor. He grabbed a fresh one and snapped it in place.

Bueller emerged from the other corner and shot at the side of the gold shack.

Antoine heard a groan. *Cashmere Coat's down.*

In the distance, another two men emerged, running.

Antoine raised his weapon. Five bursts of two. Grass kicked up. Short.

The man sheltered behind another tower.

Wha'? Pistol shot behind?! Antoine's right shoulder exploded in pain. *Oh fuck, not again!* He swung around. Two bums, guns-in-hand, were storming the sausage stand. Two bullets went somewhere to the left. Another zinged by his right cheek.

Antoine tried to lift the Heckler & Koch with his right arm, but couldn't. He couldn't feel his trigger finger. *Gotta use my bad arm.* He whipped his automatic up and fired, spraying indiscriminately. He couldn't help the barrel bouncing from left to right.

A bullet hit his Kevlar vest, jerking him to his left. *Hit Kendrey . . . blood flying . . . Shot hitting homeless man one. Can't hit 'em all!*

***

Paul had left the underground box and was now kneeling next to the long booth some fifty yards from the roller coaster platform where his bullet-riddled double "stood," silently dripping corn syrup. His back was now supported by the post. He was directly facing the sausage trailer, twenty yards away.

Michael stopped ten yards from the sausage trailer. He took the Molotov cocktail from his coat and placed it on the grass. He knelt down and lit it.

A bullet from somewhere brought another of Michael's friends down. Another bullet hit a wall next to Michael.

Paul had to get the Italian *now*. He whipped out the HK USP. He'd practiced this at least four thousand times in Baltimore, but felt his hand quivering now.

The Italian lifted his gun again, awkwardly. But his aim was effective enough to incapacitate another of Michael's friends.

Paul peered down the sight. He exhaled. *Mike is next. No. A fraction to the left.* A bead of sweat fell from his nose.

The Italian moved the barrel three inches to his right, he was pulling—Michael spun around, hit.

Paul gently squeezed the trigger. "Phutt."

Antoine's head jerked sideways. Blood splattered. He was down.

Michael grabbed his shoulder, but still managed to toss the bottle inside, turning on his heel. He fell. A tuft of grass kicked up next to his left foot.

The sausage stand exploded in bright orange flames.

Suddenly Paul's right hand jerked to the left, gun jumping out. Blood flying. *Pain.* A red spot.

Instinctively, he hit the ground. A shot screamed past, singeing the hairs on his ear. He lay flat, as he listened to gunshots from the third location.

378

In the distance, a man screamed. More gunshots. None in his direction.

Paul waited half-a-minute before turning his head towards Michael.

Michael lay on the ground, motionless. He held his shoulder, his hand red on the edges.

Paul looked further to his right. Kendrey was still down. Behind him, in the distance, three undercover cops advanced on the bumper car building, guns blazing.

But from the left side, someone shot at the two, the officer on the left went down. *They're fighting for their lives.*

Paul picked up the shell ejected from his pistol and pocketed it. He pushed himself up with his left hand to a crouching position. No shots. He ran to the blazing sausage stand. With his good hand, he threw his gun and the shell inside. His momentum carried him another five yards past it. He stopped.

A man in a black leather jacket appeared twenty yards away, crouched down and lifted an automatic rifle to his shoulder. He aimed it at Paul's chest.

Paul lifted his head to the heavens and shut his eyes. He took in a deep breath. Through the cordite and the flames, he could smell spring, the crocuses blooming, the grass. *To die on such a glorious day . . .*

A shot rang out.

His right arm jerked. *Ow pain.* Paul dropped to his knees.

But no shots from the man.

Paul looked up.

The leather-jacketed man was sprawled on the ground. He had a gaping wound on his right side.

Paul jerked his head to his right.

Kendrey, still prostrate, was holding his pistol. He raised it at Paul, then dropped it.

The sniper fire had stopped.

Paul walked to Kendrey. As he approached, he saw the forearm lift up. *He's still alive.*

This time, he had mixed feelings about seeing Kendrey in pain. On the one hand, he was glad. He hated what Kendrey had done to him. It wasn't the physical aspect. He sensed he would

379

soon recover all of his faculties and use of his right arm. It was that Kendrey had shaken his trust in the entire human race. At the same time, he couldn't help feeling some kind of bond, perverse as it was. They had spent the most harrowing four hours Paul had ever experienced, together. *For Chris sakes, he had knocked Jenkins out . . . brutally.* And now, once again, intentionally or not, Kendrey had saved him.

Still, he kicked Kendrey's pistol away before crouching down next to him.

Kendrey had been hit low on the right side. Four inches above the belt, a blood spot had grown to encompass his gut and kidneys.

"Josh. You hear me?" Paul asked.

Kendrey lifted his head up. He grinned, then let his head drop. "Yeah Paul. It hurts."

"I believe you, Josh," he said, nodding. "We'll just get you to the hotel and fix you up."

Kendrey grimaced. "I've heard that line before," he muttered through his clenched teeth.

Paul sat on his butt and laughed. "Yeah, you sure have."

"So, Paul, I'm not going to make it, am I?"

He looked at the wound. The blood was seeping out steadily, darker than it had looked in all the times he'd seen it, including November 4th. From the body diagram he remembered in high school, he concluded the bullet may have hit his liver. "Doesn't look good, Josh."

"Hurts like hell." Kendrey grimaced as he inhaled. The grimace slowly turned to a smile.

By now, police were swarming the grounds. A bird started chirping. The sound of distant sirens soon blended in with the chirping in a high-pitched duet.

"Joshua, I guessed it was you, but only just yesterday."

Kendrey smiled. "Yeah, Mr. Incarceration, what tipped you off?"

"Well, it was strange that Marty didn't stay with the company, even though they offered him a good job. He went to India instead."

"It is a wonderful place, the tourist brochures vouch for it."

"Yeah, Marty and . . . Mrs. Leboeuf? She'd fit in like a frog in a nest of herons. Come on. The new people got a card from the Missus."

"Yeah? Had she visited the Taj Mahal yet?"

"It was tear-stained and she wished she were back in the house. You know, Josh, I met Marty. He lacked the mettle to convince his wife to be quiet and take it. If Marty had orchestrated this thing, he would have had the sense to move to the U.K., which his wife would at least have tolerated."

"Pretty good there, Robinson."

"Sure. And Duprix had a little accident. Did you have anything to do with that?"

Kendrey shook his head. "Driving fast is dangerous."

"Yes, but the staff at the Lykens canteen and hardware store saw someone who looked an awful lot like you, coming in with Jenkins, whom they recognized from the photo. Said the second hunter didn't look like he could handle a gun."

"Mistaken identity."

"No, spot on. I just saw you trying to shoot me." Paul nodded. "Anyhow, I showed them your photo. They recognized you.

And before that, Josh, Latanya went to the State Department and parallel tested the machines. They were rigged, of course. And since you marked, on the side of each machine, the percent that Dooley would get, your people had to adjust the ballots."

Kendrey smiled. "Very good, Robinson. Excellent idea, wasn't it? No one watches the back-up videos. Not in Maryland four years ago, not in Ohio two years back . . ." Kendrey coughed dryly, "and not in Pennsylvania now."

"Yeah," Paul chuckled. "But we hicks did."

Kendrey smiled, then grimaced. "A fluke."

Paul shrugged his shoulders. "But why didn't you just standardize the percentage?"

"They're dumb, but they're not retarded. They'd notice."

"Makes sense. But the whole op was so . . . risky. Why not just reprogram the machines remotely."

Kendrey coughed. "Aha. The obvious. There are plenty of hackers out there that would have tried, and you know, even a few

GenXers that would have succeeded," he coughed again, "in uncovering what I had changed, if they had had access. Besides, election commissions are onto it. The voting machines are off-grid." Kendrey readjusted his weight. He grimaced. "But the old-fashioned way—nobody would have suspected. And I would've gotten away with it . . . if it hadn't been for that . . . glitch."

"C'est la vie. But you know, Josh, the question that I keep asking myself is, why didn't you kill me once you knew where I was going, or in Hollidayton, before I got to the journalist's house.

Kendrey looked up at the sky. Once again, he smiled. "Cat and mouse?" He glanced at Paul. "I knew you wouldn't be getting away. And we wanted you at the journalist's place anyway, to get the pedophile angle."

Paul scrunched his nose and shook his head.

"Robinson . . . " Kendrey continued, "Who were 'the old folks' we were going to see if you didn't get the answer from the Stoltzfusses? Your 'back-up plan.'"

Paul smiled, saying nothing for five seconds. "So why hit Jenkins?"

"Answer me first."

"No, you."

"Asshole." Kendrey coughed. "He wasn't supposed to shoot you until we had the address. He screwed up the signal."

"And he could have hit you, as we—"

"Not under normal conditions – his aim was spectacular. Without the wind, he wouldn't have missed—"Kendrey breathed in, grimacing, "But he just didn't wait. He paid for it." Kendrey smiled. "You should have seen him, hopping around with a bandage around his head for two weeks."

"He wasn't the only one. The Italian was about to end my days—"

"No, he was just toying with you, waiting until I got there."

"And when you did, you just about killed me."

Kendrey grimaced again, but the grimace broke into a smile. "Aimed for the arm, but, alas—" Kendrey coughed, pointing to Paul's arm. "I can't shoot worth a plugged nickel."

Paul shook his head. "Man, did you play me…Anyway, you should know—I released the story to the Gazette and the Times yesterday. It'll be in today's paper. . . not to mention my station."

Kendrey shrugged his shoulders. "A man's gotta do what a man's gotta do." He winced as he shifted weight.

"But here's your chance to come clean, Josh--"

"Don't take it hard, Robinson, belief systems are belief systems," Kendrey interrupted. Suddenly, he sat up, grabbing Robinson by his urine-soaked shirt. "So, Paul, who were the 'other folks'?"

He smiled. "Nobody."

"Nobody?"

"Yeah, nobody. I was just jerkin' your chain, Josh. Tit-for-tat. You wouldn't tell me about the programming, so I wouldn't tell you about Plan B."

"You dick!" Kendrey fell back down, groaning. "Promise me something . . . jerk."

"You know, you're not very well-behaved for someone who wants a favor."

"Go to my mother's grave," Kendrey said, ignoring his comment, "Greenwood cemetery in Southeast Pittsburgh. Mary Kendrey. Row fourteen in section 'C.' Twenty-fifth gravesite in. And make sure to change the flowers."

"Sorry to hear she passed—" Paul said, as solemnly as he could. "You were obviously very close . . . or were none of the calls to her?"

"Yeah, I was," Kendrey sputtered, "some years back."

"But you—"

"She's been dead for a decade."

56

Sirens blared in this semi-rural neighborhood, heretofore accustomed only to the rev of engines racing on Fridays at that time of the day. Dozens of inhabitants of the Sleepy Hollow trailer community pressed their noses against the fence to view the police swarming over the six fallen bodies.

The ambulances carried away Bueller, the pilot, still breathing, and one of the train hoppers that had made the error of standing up here.

The Harrisburg coroner's office arrived shortly thereafter for the corpses.

Within minutes, reams of yellow crime scene tape tied off what appeared to be acres of the park.

Paul looked at all the goings-on silently. He was still leaning on the shack next to the smoldering sausage stand. He found it hard to fathom that he would soon be free to live a normal life again.

Olson finally showed up. "So, Robinson, nice contraption," he started, before pulling out a handkerchief and putting it under his nose. "You'll be pleased to know the Danabold company came out five minutes ago saying that you're a modern-day American hero—all over the airwaves. You, uh, prompted them to do an internal investigation."

He lowered his head. "What? Oh, yeah, sure."

Olson nodded. Behind him were three policemen. "But you're not off the hook, yet," he said, signaling to the policemen, who promptly took some handcuffs out of their holsters. "I'm gonna have to bring you to the station for a long talk. You claim self-defense, but you stole the BMW. Olson looked at him askance. "You'll leave the room as a witness . . . or a suspect, we'll see which."

On the way to the car, Olson turned to him once again. "Before we get to the station, I'd like to know—whaddya think motivated this guy? He ended up killing twelve people, including two fine policemen."

"A sociopath with a messianic complex, hiding behind the exterior of a computer geek. Something like that. I mean, he took a bat to his colleague's head in the tunnel and could've killed me at any time that afternoon, but didn't try to do so until Hollidayton."

As they approached the cars, Paul saw Tan running towards him, her smile lighting the way, a dream.

Epilogue

After two days the stench of urine and vomit, that had penetrated into at least a couple of layers of skin, disappeared entirely. Paul's newfound taste for freedom had not. He imbibed, slowly, both the rich, sweet coffee and the reports in the papers. According to the New York Times, the company had completed an internal investigation, prompted by the discovery that their programmer, Joshua Kendrey, had collaborated with a known pedophile.

They found he had colluded with another programmer by the name of Charles Duprix, to re-profile the machines used in that county, which had also given him an excuse to visit Blair County more often than he would otherwise have been entitled to. His other accomplices included three South Philly hit men and an AWOL army ranger, as well as a military contractor, recently-deceased from natural reasons. The company further commended the brave efforts of the reporter Paul Robinson in finding the root of the problem and wished him a complete and rapid recovery, after which they would present the reward in a public ceremony.

The district attorney had dropped the pedophilia case against Paul when the tampered photos were analyzed. Furthermore, the weapon that had killed the law officers was traced to several other killings, which Fauntleroy had no problems demonstrating could not have been linked to Paul.

Finally, the state's Attorney General dismissed charges of Grand Theft Auto against him, his clean record and history of volunteer work attested to Paul's claim that he had just borrowed it in an attempt to warn the victims of impending danger.

Paul carried on scanning, not reading, whole sentences. "Danabold completes internal investigation. Foul-play by renegade programmer leads company to concur with Penn State Department . . . recommending invalidation of Blair, Berk, and Dauphin County voting results . . . remainder of machines in the state untainted, according to Department of State."

Later that day, after an afternoon of unhurried lovemaking, Paul and Latanya were once again sitting in front of the TV.

The story was the subject of a half-hour interview by Westerson Vintner. "We were wondering why results stood out so much in that county. To have such a large discrepancy in results between two adjacent Republican counties, is not entirely unusual, but rare. After meeting with Paul Robinson, a courageous reporter from WQHP, we launched an extensive internal investigation. We were shocked to find that one of our programmers had taken advantage of the trust we had placed in him, and not only broken all company policies, but like some hacker, sabotaged all our safeguards for these three counties," claimed the square-faced CEO Wilson.

"How will this affect Presidential elections in two years' time?" the interviewer asked.

"Not at all. First of all, we have re-tooled the security of the machines. They will be one hundred percent tamper-proof by the next election. One also has to place the damage into perspective. These are three counties out of sixty-seven. Results would have been different only if either Senator would have received *all* of the votes cast on *one hundred percent* of the voting machines in these three counties as well as one more, which was not the case."

"How much did the machines give to one candidate over the other?"

"We believe it was about twenty percent, for the machines affected, or less than five percent for the winning candidate in those voting districts."

"It was for candidate Dooley in all cases?"

"We cannot comment at this stage decisively, but it does appear that they gave Dooley an edge. But again, let me repeat, not enough to impact the U.S. Senate election. They may have widened his lead by a percentage point or two, that's all."

"How do you know?"

"Our statisticians have studied enough of the machines in those three districts, to conclude that one-point-seven-five to two percent, maximum, additional percentage points were accorded the winning candidate. Dooley would have still won. And the Department of State corroborates that conclusion," the spokesman said with well-rehearsed confidence. "You know, it would be a terrible thing for the country, if, because of some foul play by a disgruntled loner employee, that had no material impact on any election outcome, a new technology whose time has come would be rejected in favor of the mechanical machines proven so faulty in past elections."

Paul caressed her forearm. "You believe him?"

"Yeah, Paul," she said nodding. "Put it this way . . ." she turned, wrapping her arms around him, "I couldn't prove otherwise." She looked longingly in his eyes. "Look, we have our lives back, sweetie. Isn't that a precious enough substitute for one hundred and ten percent certainty?"

He enveloped her in his arms.

She couldn't explain why, but she felt safe, calm.

"*We also wish to reward the reporter that uncovered this crime* . . ." the spokesman continued.

They promptly sat up.

"With a fifty-thousand dollar gift to his favorite charity," the CEO responded.

Paul nodded. Latanya smiled.

"Who should it be, Latanya? The Save the African Rhino Fund or the Disabled Veterans' Society?"

"We have time to decide that, honey, the more pressing question is, who you gonna work for now, W-Q-H-P or W-Z-T-W Philadelphia?" she said, caressing his bandaged hand.

"Philly, here we come."

"Philly? . . . but you hate the Eagles. How you gonna do that?" she jibed, inwardly ecstatic at how far Paul had reached for that golden ring. And the icing on the cake was that department heads from two prominent universities from the city had told her they would be interested in interviewing her as soon as she completed her doctorate.

"I'll just suck it in and watch Pittsburgh on the sly—you know, on the iPad in the bathroom." He added, "So Tan, which charity is it going to be?"

"Should I get you a coin?"

"Naw—" Paul said, throwing his head back and looking in her eyes, "better some of your indescribably phenomenal hummus."

Acknowledgments

A sincere thanks to the Writer's Center in Bethesda and specifically Kathryn Johnson. Also a nod goes to the supportive group of American mystery and thriller writers, who not only provide gripping entertainment to readers but are also potent actors in augmenting literacy worldwide.

Thank you to Jonathan Caseley, who [reluctantly] provided me the scathing critique that prompted necessary, albeit painful, late-stage rewrites.

An enormous thank you to Caroline Thomas for a refreshing first concept edit and to Hilary Ross of New York for the second one.

Very special thanks to my wife Inguna, for both the comments and general support throughout.

And a hearty *tack* to the rest of my family, friends and contributors.

On the next page is an excerpt from GLITCH 2201, the next novel in the Robinson-Forrester series. Available May, 2015.

Risking their own lives to torment another two people didn't feel quite right. Latanya felt so conflicted, she couldn't say anything. What she wanted to express would not merely aggravate Paul, it would torture him. Slowly, like chinese finger locks. *A compromise?* "Paul, Paul. Okay."

His eyes lit up.

Latanya put her hand on his. "But listen. Let's not, let's not act prematurely. The elections aren't going to be for . . . eighteen months." Images of the fire-engulfed house, the gasping Texan and the scarfaced Parkett writhing helplessly as death took hold of him, flashed before her. She took a deep breath. "Let's plan it out. Let's make it . . . body-bag-free."

"Fine, we'll do that. You're a gem, Tan, a gem. First thing we'll do is make sure to get Kendrey's sister is out of danger."

She smiled on the inside. But on the outside, she kept her demeanor straight. She turned over Paul's hand, glancing at the light purple scar. *No conceding that point.*

****

It was 8:15 in the evening when David realized he was going to be fifteen minutes late to dinner with Elaine. But he had to finish his analysis. There was no link between the hydro-fracking, the civil engineer had assured him, and the earth tremors in north of Pittsburgh. 'Geologically impossible,' was the term he had used.

And public opinion was still strongly in favor of the drilling for natural gas. But not as strong, according to the latest poll. Another 4% drop, in fact, since December. Reasons mentioned were that while no appreciative benefits had been derived for the public schools or roads, the water quality was getting worse . . . much worse. And while more roads were being built, they were just linking the wells, not the communities. 'And jest after a road is emproved, truks terr it up," wrote another polled Appalachian.

The result was going to turn some heads. Maybe this was something Paul would be interested in reporting, now that he was covering the entire state. He would ask Elaine about that.

As the back office door quietly shut behind him, he was startled, then momentarily blinded by carlights switched to high beam.

Two men appeared out of nowhere and one of them held a crowbar, the silhouette of which David saw clearly.

He wasn't one to panic, but he knew it would be an 'unpleasant experience' as people here were fond of saying.

As the light came from the rear, he couldn't see their faces.

"Look, bud," the first one started, "You can't always believe in polls, right?"

"Well, not always," Dave responded feebly, pondering how a confrontation might affect his visa renewal chances.

"So we don't think these are quite right."

"How do you kn—"

"Just a hunch . . ."

David scratched his head. How these guys knew, he had no clue, and with each additional second he was more convinced they didn't.

"What do you want?" he asked, detecting a faint aroma of chewing tobacco in the man's breath.

"Just don't release the poll results, unless you've interviewed some people from, oh, Bradford County."

"We have, and they stand out for their satisfaction levels. If we interview more, it would skew—"

The man on the right, holding a crowbar, shrugged his shoulders. "That was a friendly suggestion."

Light glanced off a foot-long pipe to his left, suddenly flashing in front of his chest.

David felt a searing pain in his left side and a millisecond later, something snapping inside. He fell to his knees.

"We ain't gonna be so friendly next time, stranger."

Dave was overcome with nausea. Everything – the parking lot, the lights, the two strangers – began spinning, until he saw only black, then nothing.

www.ingramcontent.com/pod-product-compliance
Lightning Source LLC
Chambersburg PA
CBHW020836030726
47496CB00001B/252